L'Organisation de Coopération et de Développement Économiques (OCDE), qui a été instituée par une Convention signée le 14 décembre 1960, à Paris, a pour objectif de promouvoir des politiques visant :
— à réaliser la plus forte expansion possible de l'économie et de l'emploi et une progression du niveau de vie dans les pays Membres, tout en maintenant la stabilité financière, et contribuer ainsi au développement de l'économie mondiale ;
— à contribuer à une saine expansion économique dans les pays Membres, ainsi que non membres, en voie de développement économique ;
— à contribuer à l'expansion du commerce mondial sur une base multilatérale et non discriminatoire, conformément aux obligations internationales.

Les Membres de l'OCDE sont : la République Fédérale d'Allemagne, l'Australie, l'Autriche, la Belgique, le Canada, le Danemark, l'Espagne, les États-Unis, la Finlande, la France, la Grèce, l'Irlande, l'Islande, l'Italie, le Japon, le Luxembourg, la Norvège, la Nouvelle-Zélande, les Pays-Bas, le Portugal, le Royaume-Uni, la Suède, la Suisse et la Turquie.

L'Agence de l'OCDE pour l'Énergie Nucléaire (AEN) a été créée le 20 avril 1972, en remplacement de l'Agence Européenne pour l'Énergie Nucléaire de l'OCDE (ENEA) lors de l'adhésion du Japon à titre de Membre de plein exercice.

L'AEN groupe désormais tous les pays Membres européens de l'OCDE ainsi que l'Australie, le Canada, les États-Unis et le Japon. La Commission des Communautés Européennes participe à ses travaux.

L'AEN a pour principaux objectifs de promouvoir, entre les gouvernements qui en sont Membres, la coopération dans le domaine de la sécurité et de la réglementation nucléaires, ainsi que l'évaluation de la contribution de l'énergie nucléaire au progrès économique.

Pour atteindre ces objectifs, l'AEN :
— *encourage l'harmonisation des politiques et pratiques réglementaires dans le domaine nucléaire, en ce qui concerne notamment la sûreté des installations nucléaires, la protection de l'homme contre les radiations ionisantes et la préservation de l'environnement, la gestion des déchets radioactifs, ainsi que la responsabilité civile et les assurances en matière nucléaire ;*
— *examine régulièrement les aspects économiques et techniques de la croissance de l'énergie nucléaire et du cycle du combustible nucléaire, et évalue la demande et les capacités disponibles pour les différentes phases du cycle du combustible nucléaire, ainsi que le rôle que l'énergie nucléaire jouera dans l'avenir pour satisfaire la demande énergétique totale ;*
— *développe les échanges d'informations scientifiques et techniques concernant l'énergie nucléaire, notamment par l'intermédiaire de services communs ;*
— *met sur pied des programmes internationaux de recherche et développement, ainsi que des activités organisées et gérées en commun par les pays de l'OCDE.*

Pour ces activités, ainsi que pour d'autres travaux connexes, l'AEN collabore étroitement avec l'Agence Internationale de l'Énergie Atomique de Vienne, avec laquelle elle a conclu un Accord de coopération, ainsi qu'avec d'autres organisations internationales opérant dans le domaine nucléaire.

The ultimate objective in the management of wastes from uranium mining and milling is to dispose of them in such a manner as to provide for the protection of man and the environment from all potential effects which could possibly arise, due to the presence of both radioactive and non-radioactive contaminants in the tailings.

In recent years, much progress has been made in the management and disposal of such wastes. In particular attention has been focused on the long-term aspects, that is, on technical questions, licensing and regulatory requirements and administrative issues which must be dealt with in order to ensure that wastes which remain after completion of mining and milling operations are safely disposed of.

To review the present status of knowledge, NEA organised, with the sponsorship of the US Department of Energy, two workshops which focussed on the following long-term aspects :

1) the long-term geomorphological stability of tailings disposal sites, and the disposal facilities which are built on or in such sites ;

2) the nature, properties and potential long-term behaviour of the tailings material itself.

GEOMORPHOLOGY WORKSHOP

At the close of uranium mining and milling operations waste disposal must be completed and disposal facilities must be decommissioned and rehabilitated through engineering works that will provide for the containment of the wastes and the stability of disposal structures in the long-term. However, such facilities, depending on their siting and design, will invariably be subjected to geomorphological and climatological influences in the long-term. Hence, the principles of geomorphology should be applied to the siting, design, construction, decommissioning and rehabilitation of disposal facilities to provide for long-term containment and stability of tailings. The geomorphological workshop reviewed the state-of-the-art in this area, illustrated the application of the principles of geomorphology to the evaluation and selection of stable sites, and provided guidance on the design and evaluation of the long-term performance and stability of disposal facilities.

TAILINGS WORKSHOP

The nature and properties of tailings and their behaviour after disposal influence the potential impacts which might occur in the long-term. This Workshop reviewed technologies for uranium ore processing and tailings conditioning with a view to identifying what improvements could be made in the characteristics of tailings in order to enhance the long-term safety of their disposal.

La gestion des résidus de l'extraction et du traitement du minerai d'uranium a pour objectif ultime d'assurer l'évacuation matérielle de ces déchets de telle manière que l'homme et l'environnement soient protégés des effets qui pourraient découler de la présence, dans ces déchets, de contaminants aussi bien radioactifs que non radioactifs.

Ces dernières années, d'importants progrès ont été réalisés en matière de gestion et d'évacuation de ces déchets. L'attention s'est notamment portée sur les aspects à long terme, c'est-à-dire sur les problèmes techniques, les obligations en matière d'autorisation, les prescriptions réglementaires et les questions administratives auxquels des solutions doivent être apportées pour que les déchets qui restent après l'achèvement des opérations d'extraction et de traitement soient évacués en toute sécurité.

Afin de faire un tour d'horizon des connaissances actuelles, l'AEN a organisé, avec le parrainage du Ministère de l'Energie des Etats-Unis, deux réunions de travail qui étaient consacrées aux aspects à long terme suivants :

1) stabilité géomorphologique à long terme des sites d'évacuation de résidus et des installations d'évacuation qui sont construites à la surface ou dans le sous-sol de ces sites ;

2) nature, propriétés et comportement potentiel à long terme des déchets eux-mêmes.

REUNION DE TRAVAIL SUR LA GEOMORPHOLOGIE

Lorsque cessent les opérations d'extraction et de traitement du minerai d'uranium, il faut évacuer tous les déchets, déclasser les installations d'évacuation et remettre en état le site en procédant à des travaux techniques qui assureront de façon durable le confinement des déchets et la stabilité des structures d'évacuation. Toutefois, ces installations, suivant leur implantation et leur conception, seront invariablement soumises à des effets géomorphologiques et climatologiques à long terme. Il conviendrait donc d'appliquer les principes de la géomorphologie au choix des sites, à la conception, à la construction et au déclassement des installations d'évacuation, ainsi qu'à la remise en état des sites afin d'assurer le confinement et la stabilité à long terme des résidus. Les participants à la réunion de travail sur la géomorphologie ont fait un tour d'horizon de l'état des connaissances dans ce domaine, donné des exemples d'application des principes de la géomorphologie à l'évaluation et au choix de sites stables, et fourni des orientations sur la conception des installations d'évacuation et l'évaluation de leur efficacité et de leur stabilité à long terme.

REUNION DE TRAVAIL SUR LES RESIDUS

La nature et les propriétés des résidus, ainsi que leur comportement après évacuation, influent sur les éventuelles incidences à long terme. Les participants à cette réunion de travail ont analysé les techniques de traitement du minerai et de conditionnement des résidus en vue d'identifier les améliorations qui pourraient être apportées aux caractéristiques des résidus afin d'accroître la sûreté à long terme des méthodes d'évacuation.

GEOMORPHOLOGICAL EVALUATION
OF THE LONG-TERM STABILITY
OF URANIUM MILL TAILINGS DISPOSAL SITES

RÉUNION DE TRAVAIL SUR

L'EVALUATION GEOMORPHOLOGIQUE DE
LA STABILITE A LONG TERME
DES SITES D'EVACUATION DES RESIDUS
DU TRAITEMENT DE L'URANIUM

TABLE OF CONTENTS
TABLE DES MATIERES

BACKGROUND TO THE NUCLEAR ENERGY AGENCY PROGRAMME ON THE
LONG-TERM ASPECTS OF THE MANAGEMENT AND DISPOSAL OF
URANIUM MILL TAILINGS
P.J. Rafferty (NEA) 11

HILLSLOPE AND SCARP MORPHOLOGY
T. Toy (United States) 19

DESCRIPTION OF THE PANEL MINE TAILINGS AREA
P.F. Pullen, J.B. Davis (Canada) 29

THE GEOMORPHOLOGY OF THE WET AND DRY TROPICS AND PROBLEMS
ASSOCIATED WITH THE STORAGE OF URANIUM TAILINGS IN NORTHERN
AUSTRALIA
R.F. Warner, G. Pickup (Australia) 45

GEOMORPHIC ASSESSMENT OF URANIUM MILL TAILINGS DISPOSAL SITES
S.A. Schumm, J.E. Costa, T. Toy, J. Knox (United States)
R. Warner (Australia)
J. Scott (Canada) 69

BIBLIOGRAPHY OF SELECTED REFERENCES
BIBLIOGRAPHIE : REFERENCES SELECTIONNEES 81

LIST OF PARTICIPANTS - LISTE DES PARTICIPANTS 89

BACKGROUND TO THE NUCLEAR ENERGY AGENCY
PROGRAMME ON THE LONG-TERM ASPECTS OF
THE MANAGEMENT AND DISPOSAL OF
URANIUM MILL TAILINGS

P.J. Rafferty

Radiation Protection and Waste Management Division
OECD-Nuclear Energy Agency
Paris, France

ABSTRACT

        This paper provides a summary of the background and the
terms of reference of the Nuclear Energy Agency Programme on the
long-term aspects of the management and disposal of uranium mill
tailings. In particular the paper discusses various background
perspectives to the Workshop on Geomorphological Evaluation of the
Long-Term Stability of Uranium Mill Tailings Disposal Sites.

INTRODUCTION

As part of its ongoing programme concerning radiation protection and radioactive waste management associated with uranium mining and milling, the Nuclear Energy Agency organised a Seminar on "Management, Stabilisation and Environmental Impact of Uranium Mill Tailings" in Albuquerque, New Mexico, USA in July 1978. Following that Seminar, the Agency in 1979 commenced a programme of work dealing specifically with the long-term aspects of uranium mill tailings management and disposal, under the joint sponsorship of the Committee on Radiation Protection and Public Health and the Radioactive Waste Management Committee. A Co-ordinating Group on the Management of Uranium Mill Tailings (CGUMT), composed of representatives from ten countries, was formed to oversee the programme.

The programme has been carried out by three working groups respectively on radiological protection aspects, engineering aspects and environmental monitoring. The terms of reference of these groups are given in Annex 1.

The primary objective of the study has been to examine the long-term aspects of uranium mill tailings management and disposal in an attempt to apply the ICRP System of dose limitation, in particular the principle of optimisation of radiation protection, in decision-making and physical actions associated with the long-term aspects of the management and disposal of wastes from uranium mining and milling.

In parallel the study has also examined engineering aspects of tailings disposal, focussing on technologies for the management and disposal of tailings, on the design, construction, operation and decommissioning of tailings disposal facilities, and in particular on physical performance objectives and criteria for meeting engineering requirements for long-term containment and stability of waste disposal systems and long-term environment protection and land use goals. In this part of the study the role of long-term geomorphological and climatological actions on the stability and integrity of tailings disposal structures has been examined further. Site-specific engineering requirements which are needed to withstand these actions play the dominant role in engineering for long-term containment and stability of tailings disposal systems.

The Co-ordinating Group, in conjunction with the Working Groups, recommended that a Workshop be held to examine geomorphologic influences on the long-term stability of uranium mill tailings disposal facilities. The Workshop was sponsored by the Nuclear Energy Agency and the US Department of Energy, and was hosted by the US Department of Energy through Sandia National Laboratories at Colorado State University, Colorado, USA on October 28 - 30, 1981.

Further background to the need for, the objectives and scope of the Workshop are provided below. The paper also provides a number of perspectives of the stability of uranium mill tailings disposal sites subject to long-term geomorphological influences. The time frame for which an assurance of the stability of a disposal facility is required is such that new approaches to the engineering of such facilities need to be considered. This, essentially, was the objective of the Workshop.

TECHNICAL BACKGROUND FOR THE WORKSHOP

The overall objective of management and disposal of wastes from uranium mining and milling is to physically dispose of the wastes in such a way as to provide for the protection of man and the environment from radiological and non-radiological hazards which could possibly arise in the long-term.

It has been recognised that past practices of uranium mill tailings management, in terms of radiological protection and radioactive waste management principles, have in many cases been inadequate. However, it is now considered that relatively secure emplacement, i.e. permanent disposal, of tailings is not an insurmountable problem and that various feasible engineering alternatives exist which can be regarded as adequate and acceptable for long-term containment. It is generally accepted that if the best practicable engineering technology which is currently available is used, no gross or unacceptable environmental impact can be expected to result.

The persistence of radioactivity within the tailings requires that the period of time for which tailings should be isolated from the environment should effectively be indefinite, and that engineering solutions for tailings disposal should ideally be permanent and should also ideally provide for a very high degree of containment.

However, it is also recognised that even the most soundly engineered tailings containment facility cannot be expected to provide absolute containment of all the contaminants in the tailings. Even when a well designed tailings disposal facility is finally completed and decommissioned, i.e. when the containment system can be expected to perform to the most stringent design performance standards, some small but finite quantity of contaminants can be expected to escape through the containment system. Nevertheless, the engineering of the containment system must attempt to ensure that radiological protection principles (i.e. the ICRP System of Dose Limitation) are observed and that the containment system complies, throughout its design life, with design performance standards and authorised release limits which are set for the decommissioning and rehabilitation of the containment system.

Engineering alternatives for tailings containment systems, such as are discussed in the IAEA Technical Report N° 209/17 are representative of the best practicable technology which is currently available. These solutions, depending on their design bases, can be expected to provide maximum containment of the mobile contaminants within the tailings (by maintaining releases of contaminants through the containment system within design limits) and can also be expected to provide complete and secure retention of the solid tailings material itself for the "design life" of the containment structure. Needless to say, the most desirable alternative for disposal of tailings is more often than not to place them below grade, either in the mine from whence they came or in specially excavated below grade pits.

The "design life" is a concept of the engineering design process and typically can be expected to be of the order of one hundred to several hundred years for a conventional approach to the engineering design and construction of a water storage dam or a waste disposal facility. However, an engineered structure or containment system may well and, in fact, often does last considerably longer than its "design life". In the case of containment facilities for the disposal of uranium mill tailings, the desirable objective is to provide a system that will retain integrity for the indefinite future.

However, it is recognised that even the most soundly engineered containment system cannot be expected to ensure permanent retention in the indefinite future of the bulk solid tailings material itself. Due largely to geomorphological and climatological processes which can be expected to occur in the long-term, it can be expected that a containment system may eventually suffer partial or substantial damage, consequently leading to the dispersal of tailings material itself.

In this context, it is appropriate to recall the conclusions of Nelson and Shepherd /27 after a study of the long-term stability of uranium mill tailings disposal alternatives following decommissioning, rehabilitation and abandonment. Three long-term periods were considered; these were the short long-term (several hundred years), medium long-term (several thousand years) and long-long-term (100,000 years; by which time the activity of thorium-230 and subsequent radionuclides in the decay chain are approaching secular equilibrium with the residual uranium-238 activity in the tailings). Within short long-term periods, natural processes will have a small effect and failures that might occur would be design related. Thus for short long-term periods the selection of the site and the engineering design of the containment system will govern its performance. For the long-term period natural geomorphological processes would predominate. The medium long-term period represents a transition from the engineering dominance to the geomorphological dominance.

However, it should be noted that long-term geomorphological processes do not and need not necessarily always result in the loss of integrity of a containment system, for example, the case of a well engineered below grade facility constructed in a stable landform. The integrity of a containment system, and its performance, may, in time, either increase, be maintained, or decrease, depending on the siting and design of the disposal facility and the specific geomorphological environment and climatological processes; for example, in certain circumstances, geomorphological processes (e.g. subsidence or depositional processes) may result in an increase in the physical containment of the tailings rather than a decrease.

Hence it is evident that, in cases where geomorphological processes lead to a decrease of integrity of the containment system, the siting and structural design of a tailings disposal facility govern:

.   the performance of the facility in containing gaseous and waterborne contaminants;

.   the rate at which the facility begins to erode, once erosion begins, and

.   the length of time for which the structure retains integrity, i.e. until the release of tailings material begins.

The most advanced engineering design and construction practice can, and often does, incorporate qualitative consideration of the effect of predictable geomorphological and climatological processes on the performance of a containment system and on the integrity of the system itself in the long-term. The period over which such processes can be more or less predicted with any reliability may typically be of the order of a few thousand years (1,000 to 10,000 years). During such periods it should be possible to expect that a well designed containment system will remain substantially unaffected by these processes.

However, the difficulty of being unable to predict either quantitatively or qualitatively the nature and extent of all manner of geomorphological and climatological processes in the long-term, i.e. beyond the first few thousand years, makes it difficult to be confident that the release of contaminants will not exceed design performance standards or authorised limites, or that the containment system will retain its integrity. Over these time scales extreme events, such as extremely large earthquakes and floods, whose magnitude could exceed the design bases for such events, could also be expected to occur. These would more probably result in partial or substantial breaching of the containment system and consequent dispersal of tailings material. The engineering implications for the integrity of the containment system and the radiological consequences of such failures cannot be satisfactorily assessed within current perspectives. However, because of the low specific activity of the tailings, the consequences of partial or substantial breaching of the containment system and consequent dispersal of tailings would not lead to catastrophic or necessarily even significant radiological impact, since long and sustained exposure to radioactivity from tailings would be required to produce detectable adverse results.

It is evident that by making use of the best practicable engineering technology which is currently available, the continuing radiological impact which might result in man in the long-term as a result of the escape and dispersal of radioactive contaminants from uranium mill tailings, whether containment facilities ultimately retain integrity or not, is most likely to be relatively small even when compared with natural background radiation. Over the time scales, such as are discussed above, significant sociological, technological, climatological and geomorphological changes can be expected to occur. Since the nature and extent of these changes cannot be anticipated today, our responsability to future generations requires that uranium mill tailings be securely contained and stabilised for an indefinite period, using the best practicable technology currently available. Thus, the design and construction of facilities should be optimised in a sound engineering manner within realistic and practical perspectives of risks, costs and benefits.

Tailings disposal containment systems should depend for their primary integrity on substantial physical barriers to prevent the escape of contaminants through the containment system and to prevent the dispersal of tailings material itself. The design of the physical barriers should ideally reproduce and possibly improve on, if possible, those characteristics of natural materials and formations which retain or retard the transport of contaminants or tailings into the environment.

The siting and design of the containment system should also provide for the action on the system of expected or predictable geological, geomorphological, hydrogeological and climatological forces in the long-term. These forces and scenarios should be thoroughly analysed in the planning and design stage in order to circumscribe the uncertainties about long-term stability. Such studies require thorough collection, analysis and application of geological, geomorphological, hydrogeological, geographical and climatological data in the region of the site of interest. The final design should provide for the ability to withstand such forces for a prescribed time period, or should be such that the action of such forces in time increases the integrity of the containment system rather than decreasing it.

The disposal system should ideally be designed so that, following the completion of decommissioning and rehabilitation works, no active ongoing care and maintenance is required to ensure long-term stability, beyond a reasonable initial period.

Against this background, it was therefore the purpose of the Workshop to focus on various areas of concern in the engineering of disposal facilities for uranium mill tailings to ensure containment performance and structural integrity in the long-term. The areas of concern on which the Workshop concentrated were threefold, and these therefore constituted the specific objectives of the Workshop, which were as follows:-

₀ <u>To examine geomorphologic controls on site stability</u>

The Workshop examined the fundamentals of geomorphology
in order to understand how these influence the long-term stability
of landforms and the forces which, in the long-term, will affect
any containment structure which is built on or in such landforms.

. <u>To demonstrate the application of principles of
geomorphology to the siting and design of stable
tailings disposal containment systems</u>

The Workshop demonstrated how knowledge of the
geomorphology of the locality for a uranium mill tailings disposal
facility can be applied in the planning and design process
(to siting, design, construction, decommissioning and rehabilitation)
for long-term stability.

. <u>To develop a procedure for the evaluation of long-term
stability of tailings disposal sites</u>

In view of the need to have confidence in the stability
of engineered tailings disposal structures in the long-term
(in terms of many thousands of years), the Workshop attempted to
outline a planning and design procedure for the evaluation of the
long-term performance and stability of such structures and the
sites on or in which they are built.

<u>REFERENCES</u>

/1̶7 IAEA Technical Report N° 209
     "Current Practices and Options for Confinement of
     Uranium Mill Tailings"
     IAEA, Vienna 1981.

/2̶7 Nelson, John D; Shepherd, Thomas A (1978)
     for Argonne National Laboratories
     Final Report, "Evaluation of Long-Term Stability of
     Uranium Mill Tailings Disposal Alternatives"
     April 1978 (337 pp)
     Geotechnical Engineering Program
     Civil Engineering Department
     Colorado State University

     Shepherd, Thomas A; Nelson, John D (1978)
     "Long-Term Stability of Uranium Mill Tailings";
     an excellent summary of the above study.
     Proceedings of (First Annual) Symposium on
     Uranium Mill Tailings Management (Volume I, pp 155-172)
     Colorado State University
     November 20, 21 1978.

TERMS OF REFERENCE OF THE NEA PROGRAMME ON THE LONG-TERM ASPECTS OF THE MANAGEMENT AND DISPOSAL OF URANIUM MILL TAILINGS

Terms of Reference of Working Group 1

1. To apply the ICRP System of dose limitation to the development of radioactive waste management principles for specific application to the long-term management and disposal of wastes from uranium mining and milling and, in particular, to develop long-term radiological protection performance objectives and criteria for retention facilities for management and long-term disposal of these wastes.

2. To develop appropriate methods for modelling the release of contaminants from wastes from uranium mining and milling, their dispersion in the environment by atmospheric and aquatic pathways, and the assessment of their impact in the long-term.

3. To examine the feasibility and develop, if possible, a methodology for comparative evaluation of, and for optimisation of radiological protection provided by alternative types of retention facilities for the management and long-term disposal of wastes from uranium mining and milling.

4. To illustrate the application of the above optimisation methodology with quantitative examples.

5. To consider the comparative impact of non-radiological contaminants, to develop, if possible, an equivalent methodology for evaluation and optimisation of environmental protection, and to develop environmental protection performance objectives and criteria associated with non-radiological contaminants.

6. To consider carefully and identify problem areas still requiring further research and development effort and to suggest mechanisms for co-ooperative international activities in these areas.

Terms of Reference of Working Group 2

1. To critically review current and developing technologies for the management and disposal of all wastes from uranium mining and milling, and covering all aspects of production, processing, preparation and placement of such wastes which are relevant to their long-term management.

2. In view of the above critical review, to study and recommend what remains to be done in the design, construction, operation, decommissioning and rehabilitation of waste retention and disposal facilities, using alternative technical solutions, for the long-term management of wastes from uranium mining and milling, having in mind the performance objectives and criteria which are to be developed by Working Group 1.

3. To provide technical inputs to the work of Working Group 1 for the purposes of illustrating the optimisation methodology with quantitative examples.

4.   To consider carefully and identify problem areas still requiring
     further research and development effort, and suggest mechanisms
     for co-operative international activities in these areas.

Terms of Reference of Working Group 3

1.   To review the different types of environmental monitoring
     (pre-operational baseline data gathering; operational sur-
     veillance; post-operational surveillance; environmental model
     validation; research; etc.) with a view to:

     (a) identifying which types and features of environmental
         monitoring programmes are relevant to the assessment of
         radiological and non-radiological environmental impact
         associated with wastes from uranium mining and milling,
         particularly in the long-term;

     (b) identifying the gaps which still exist in knowledge of the
         design, development and operation of environmental monitor-
         ing programmes, particularly those associated with long-
         term post-operational monitoring.

2.   To consider and make recommendations on the scientific aspects
     of the design and operation of, and to prepare a guideline on
     the requirements for and techniques of environmental monitoring
     programmes which are relevant to the assessment of environ-
     mental impact associated with wastes from uranium mining and
     milling, particularly in the long-term.

3.   To provide specific expertise and data input to Working Group 1
     for the development of environmental impact assessment models
     which are required for the application of the optimisation
     methodology.

HILLSLOPE AND SCARP MORPHOLOGY

Dr. T. Toy

Associate Professor
Department of Geography
University of Denver, Colorado, USA

ABSTRACT

In this paper some common processes in hillslope and scarp morphology, their rates of operation in various climatic regions and some of the factors affecting their stability are discussed.

INTRODUCTION

The terrestrial portion of the earth's surface can be divided into two broad landform categories, hillslopes and channels. All landforms are generally considered to be the product of geomorphic processes operating through time within a given environmental setting. In this paper, some common hillslope processes, their rates of operation in various climatic regions, and some of the factors affecting their stability are described.

Hillslopes may be variously classified. One approach focuses upon profile geometry. Thus, hillslopes may be described as convex-upward, concave-upward, convex-concave or they may be complex, composed of multiple segments or elements of different combinations of attitudes and curvatures. Climate, as it affects geomorphic processes, and geology, are probably the two most important components of the environmental setting which influence geomorphic form.

Climatic conditions have changed through time and thus it is possible that hillslope form, at least in part, is a consequence of past climatic processes. Accordingly, hillslopes could be categorized as (a) relict -with form being the product of past processes; (b) equilibrium -with form being the product of present processes; or (c) composite - with form being the product of both past and present processes. In the case of reclaimed hillslopes, most of the influences of past processes have been eradicated by, for example, the disturbances of mining. It is probably permissible, therefore, to focus upon present processes while recognizing that there may be changes in climate and processes in the future.

HILLSLOPE PROCESSES

Generally, hillslopes are subjected to two principal types of processes, erosion and mass-movement. Erosion refers to the entrainment and transportation of earth materials by water, wind, or ice. Except in local areas, water is usually regarded as the dominant erosion agent. Extensive research has identified the role of various components of the environmental setting in determining erosion rates. Sheetwash and, perhaps, rill erosion are largely a function of precipitation energy, slope angle and length, inherent soil erodibility, vegetation cover, and surface management practices. Estimates of erosion rates may be obtained using the Universal Soil Loss Equation which is based upon the aforementioned variables. However, it may be more useful to select an appropriate erosion rate and solve the equation for the value of surface management practices which would be necessary to achieve this erosion rate. If a maintenance-free hillslope system is the goal, then it may be more practical to adjust other variables, such as slope angle and length or vegetation cover, to attain the desired erosion rate on reclaimed hillslopes.

Natural erosion rates may differ considerably from place to place (Table 1). If it is assumed that the environmental setting will remain constant through time, it is possible to extend these rates into the future. Accordingly, hillslope erosion may range from 100 m to 0.1 m per 10,000 years. Sediment yield research suggests that erosion rates are highest in semiarid climatic regions and decrease toward more arid and more humid regions. Human activities which remove vegetation, alter soil properties, or increase slope angle and length may increase erosion rates by an order of magnitude or more, in extreme cases.

Mass-movement is a generic term encompassing the suite of gravity-driven downslope movement of earth materials. The type and rate of movement may vary considerably depending on climatic conditions and geologic configurations. Some types seem

Table 1

NATURAL EROSION RATES

	mm/yr	m/1,000 yrs	m/10,000 yrs
Badland, N.D.	3.6–10.4	10	100
Temperate U.S.	.01–.06	.01	.1

to occur through a wide range of climates, although the rates may be substantially different (soil creep). Other types are restricted to certain climatic regions (solifluction).

To illustrate the variability in rates, discussion here is focussed upon one of the more ubiquitous mass–movement types, soil creep.

Natural soil creep rates (Table 2) may differ considerably from place to place. Again, assuming that the environmental setting will remain constant through time, these rates may be extended into the future. Accordingly, soil creep may range from 120 m to 10 m per 10,000 years. Usually, the top decimeter of the soil is most affected.

Prediction of mass–movement rates is rather difficult. Often, apparent causes are no more than triggers producing movement of hillslopes on the threshold of stability. The difficulties arise from the recognition of inherent instability and the prediction of effective triggering events or conditions. Frequently, the terrain must be examined for physical evidence of past mass–movements and, if such evidence is present, it may be inferred that future movements are at least a possibility.

There is ample evidence of man's capacity to initiate mass–movements. Usually this results from increasing stress on the hillslope system, decreasing strength, or providing a trigger, such as the vibrations from blasting operations.

Table 2

SOIL CREEP RATES

	mm/yr	m/1,000 yr	m/10,000 yr
Semiarid	5 – 12	12	120
Temperate	1 – 2	1	10

SCARPS

Scarps, cliffs or free–face slopes are particular hillslope types characterized by a nearly vertical element and, often, a talus or scree slope of unconsolidated material formed at the base. Scarps may be caused by tectonic activity or through normal weathering and erosion processes operating on particular geologic configurations. Based upon geologic configuration and its influence on form, scarps may be categorized as (a) <u>simple</u> -developed on one rock type; (b) <u>compound</u> -developed on two rock types; or (c) <u>complex</u> -developed on more than two differing rock types. Frequently, the geologic strata involved are nearly horizontal in attitude.

Rock-fall is a common geomorphic process on scarps. Cracks develop in the strata forming the free-face. These cracks are enlarged through time by ice wedging, cleft-water pressure in joints, plant root growth, mineral hydration, or as a consequence of basal support removal. Eventually, the cloven rock mass separates from the parent strata and plummets to the slope below.

Research results provide some notion of scarp retreat rates under certain conditions (Table 3). It is apparent that rock type is a significant factor affecting rates; soft shales may retreat 10 to 1,000 times faster than harder rock types. Climate appears to be another noteworthy factor; hard rock types in humid climates appear to retreat faster than similar rock types in arid regions. It should be recognized that these average retreat rates mask considerable short-term variabilities. Rock-fall is a rather sporadic type of mass-movement.

<u>Table 3</u>

RATES OF SCARP RETREAT

(from Young, 1972)

Rock Type	Climate	Recession mm/yr
Shale	Semiarid	2 - 13
Granite, gneiss	Rainforest	2 - 20
Sandstone	Humid temperate	0.5
Sandstone	Semiarid/arid	0.6
Sandstone	Semiarid/arid	0.2
Sandstone	Arid	0.00004

As mentioned above, mass-movement prediction, including rock-fall, is rather difficult. However, some generalizations are possible. Rock-falls tend to occur with maximum frequency in the autumn and spring. This pattern has been used to support ice-wedging as a triggering mechanism because freeze-thaw cycles are most frequent during these seasons. If basal support removal by stream erosion is the operational trigger, then rock-fall may be associated with the season of maximum stream discharge.

Once again, human activities which increase the stress on the hillslope system, decrease system strength or augment the crack enlargement process, may increase the frequency of rock-fall and the rate of scarp retreat. Among a wide variety of possible human actions having these affects are irrigating the caprock at the top of a scarp, repositioning a stream to the base of the scarp, increasing the discharge of a stream located beneath the scarp, or creating vibrations through blasting operations.

HILLSLOPE AND SCARP STABILITY

Stability is a state in which slight perturbations of the variables defining the system do not lead to a progression to a new

state. Based on this definition, it can be inferred that stable
hillslopes and scarps are those which endure minor changes of the
environmental setting without substantive changes in position or
morphometry. The term "stability" usually refers only to the
absence of mass-movement on the hillslope or scarp. However, the
absence of accelerated erosion on hillslopes can be included as an
additional indicator of stability on these landforms.

        In assessing the stability of hillslopes and scarps at
uranium tailing disposal sites, the spatial and temporal limits of
the evaluation must first be identified. Spatially, both the stability of hillslopes and scarps must be considered at the disposal
site itself and the stability of those in the surrounding region.
Because a landscape is composed of interrelated landforms, instability of hillslopes or scarps in proximity to the disposal site may
lead to site instability. Additionally, the stability of hillslopes existing before tailings disposal at a site must be considered as well as those slopes created by the disposal process. Pre-
disposal hillslopes provide the site foundation. Post-disposal
hillslopes may be inherently unstable unless measures are taken to
balance the various stresses and strengths.

        Geomorphologists often consider a hillslope to possess
long-term stability if it is not subject to mass-movement within
a hundred years or so. However, it is conjectured that hillslopes
associated with uranium tailing disposal sites should remain stable
for one to ten thousand years because of the need for more permanent
containment.

        Whereas one may be rather confident of a stability evalua-
tion covering the one hundred-year period, one must be somewhat
more skeptical of an evaluation pertaining to thousands of years.
Environmental conditions may vary considerably during that period.
Furthermore, rock weathering, and its effect on material strength,
is cumulative through time. Nevertheless, it is quite possible to
eliminate the more geomorphically hazardous sites based on reevalua-
tion of present conditions. It may also be possible to identify
sites which have been stable for very long time periods in a particular locale.

        A wide variety of factors contribute to the stability or
instability of hillslopes and scarps. Table 4 contains a list of
site factors and characteristics which could be used in stability
evaluations. Two categories of characteristics are included: (1)
those which influence the rates of geomorphic processes operating at
a site, and (2) features revealing the work performed by past
processes. The "primary variables" included in this table are those
that are likely to be most influential in determining stability.
The "secondary variables" may be important in certain situations,
or are rather indirectly related to stability.

        It is unlikely that data will be available for all
variables at a given site. While complete information would be
desirable, it is probably not necessary. Several variables, indi-
vidually or taken together, may lead to the same conclusion regarding stability.

CONCLUSION

        Each hillslope and scarp is subjected to a variety of
geomorphic processes. Process rates are determined by site conditions which can be readily modified by human activities. Thus,
the difference between natural and disturbed sites is largely the
result of differences in process rates rather than differences in the
processes themselves.

Lastly, it should be recognized that hillslopes and scarps cannot be disassociated from the rest of the landscape. Hillslope and channel development are interrelated. Modification of one is likely to produce a response in the other.

Table 4

SITE FACTORS AND CHARACTERISTICS RELEVANT
TO STABILITY EVALUATION

I. Factors Related to Hillslope Erosion

   A. Climate

      1. Precipitation
         a. Primary variables
            1) precipitation form (rain, snow)
            2) rainfall intensity ⎫
            3) rainfall duration ⎬ energy
         b. Secondary variable
            1) seasonality
      2. Temperature
         a. Primary--none
         b. Secondary
            1) freeze-thaw cycles
            2) average temperature (PET)

   B. Soils

      1. Physical properties
         a. Primary variables
            1) particle-size distribution ⎫
            2) organic matter content ⎬ erodibility (K)
            3) infiltration capacity
          b. Secondary variables
            1) structure
            2) color
            3) profile development
      2. Chemical properties
         a. Primary variable
            1) clay type (montmorillonite, etc.)
         b. Secondary variables
            1) electrical conductivity
            2) exchangeable sodium percent

   C. Topography

      1. Slope angle
         a. Primary variables
            1) percent inclination
            2) aspect
         b. Secondary variable
            1) general form--convexity (knob, spur) or
              concavity (hollow)
      2. Slope length
         a. Primary variable
            1) field length
         b. Secondary variable--none

3. Microtopography
    a. Primary variables
        1) rill development
        2) gully formation
        3) stone armoring
        4) piping
    b. Secondary variable
        1) dessication cracks

D. Vegetation

    1. Density
        a. Primary variables
            1) crown cover
            2) ground cover
        b. Secondary variables--none
    2. Type
        a. Primary variables
            1) species composition--grasses, forbs, trees, etc.
            2) durability--annuals, perennials
        b. Secondary variables--none
    3. Condition
        a. Primary variables
            1) general health
            2) abrupt changes in density
        b. Secondary variable
            1) root exposure

E. Land use

    1. Type
        a. Primary variables
            1) agriculture
            2) grazing
            3) cultural impacts--residential, recreational
        b. Secondary variable
            1) seasonality of use
    2. Intensity of use
        a. Primary variables
            1) agricultural cropping system
            2) grazing density: carrying capacity/utilization
            3) urbanization
        b. Secondary variable
            1) seasonality

II. Factors Related to Mass-Movement (Including Rock-Fall)

    A. Climate

        1. Precipitation
            a. Primary variables
                1) precipitation form
                2) rainfall intensity
                3) rainfall duration
                4) seasonality

            b. Secondary variables--none
        2. Temperature
            a. Primary variables
                1) freeze-thaw cycles
                2) average temperature (chemical weathering)
            b. Secondary variable
                1) diurnal range (if one accepts thermal expansion
                   and contraction as functional)

B. Soils

    1. Physical composition
       a. Primary variables
          1) particle-size distribution
          2) infiltration capacity
          3) profile development (including depth)
       b. Secondary variable
          1) soil structure
    2. Chemical properties
       a. Primary variable
          1) clay type
       b. Secondary variables--none
    3. Strength properties
       a. Primary variables
          1) shear strength
          2) liquidity index
          3) arrangement of competent/incompetent horizons
          4) plasticity
       b. Secondary variables--none

C. Topography

    1. Slope angle
       a. Primary variables
          1) percent inclination
          2) aspect
       b. Secondary variable
          1) general form--convexity, concavity
    2. Slope length
       a. Primary variable
          1) field length
       b. Secondary variables--none
    3. Macrotopographic
       a. Primary variables
          1) regional relief
          2) local relief
       b. Secondary variables--none
    4. Microtopographic
       a. Primary variables
          1) hummocky topography
          2) scars
          3) lobate form
          4) unconsolidated debris
          5) canted physical or cultural features
       b. Secondary variables--none

D. Geology

    1. Lithology
       a. Primary variables
          1) rock type
          2) cement strength (sedimentary rock)
       b. Secondary variable
          1) facies change
    2. Structure
       a. Primary variables
          1) dip direction
          2) joint density
          3) faults
       b. Secondary variables
          1) foliation (metamorphic rocks)
          2) gneissic layering (metamorphic rocks)

3. Strength properties
   a. Primary variables
      1) extent of weathering
      2) depth of weathering
      3) compressive strength
      4) arrangement of competent/incompetent beds
   b. Secondary variable
      1) weathering products
4. Tectonic history

E. Hydrology

1. Ground water
   a. Primary variables
      1) phreatic line and seasonal fluctuation
      2) pore-water pressure
      3) seepage lines
   b. Secondary variable
      1) ground water chemistry
2. Surface water
   a. Primary variables
      1) surface water ponding
      2) channel position relative to slopes
      3) evidence of incision, deposition
      4) evidence of channel migration
      5) flow regime and seasonal fluctuation
      6) knickpoint location
      7) channel metamorphosis
   b. Secondary variables--none

F. Vegetation

1. Density
   a. Primary variables
      1) crown cover
      2) ground cover
   b. Secondary variables--none
2. Pattern
   a. Primary variables
      1) abrupt changes in density or type
      2) rafting
      3) canted trees
      4) abrupt changes in age
   b. Secondary variable
      1) alignment of dead vegetation (?)

DESCRIPTION OF THE PANEL MINE TAILINGS AREA
RIO ALGOM LTD., ELLIOT LAKE, ONTARIO, CANADA

P.F. Pullen, Consultant
J.B. Davis, Golders Associates

Rio Algom Ltd.,
Elliot Lake, Ontario, Canada

ABSTRACT

The geomorphological setting of the Panel Mine tailings management area is described in relation to the geology of the Canadian Shield subject to recent glaciation and subsequent weathering and erosion in a humid climate with temperature extremes. For the deposition of uranium tailings a topographic low is chosen that is surrounded by bedrock with a low water permeability. This latter is evaluated by a detailed geological investigation and by drilling to investigate the relative permeability of suspected seepage paths. It is estimated that seepage from the basin used will be less than one litre per second. Monitor wells have been established to determine the quality of the groundwater flows.

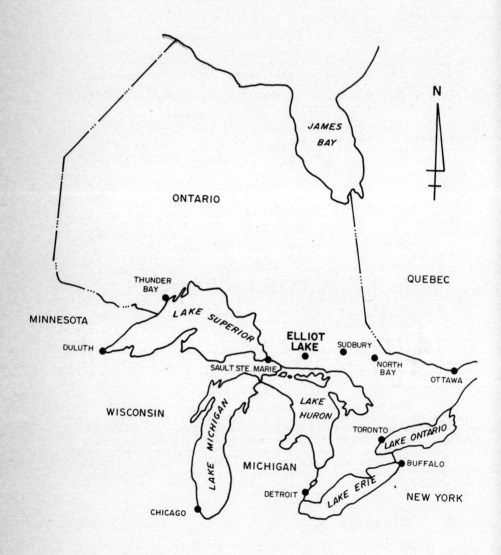

Figure 1. SITE LOCATION PLAN

INTRODUCTION

The Elliot Lake uranium mining area is situated within a block of nine townships that form the town of Elliot Lake, Ontario, just north of Lake Huron and midway between Sudbury and Sault Ste Marie as shown in Figure 1. Within the period 1955-1958 twelve mines were brought into production in the area given in Figure 2. Because the United States Atomic Energy Commission would not extend the contracts for uranium, most of the mines and mills closed down in the early 1960's, including the Panel mine. As a result the population of the Elliot Lake area dropped from over 25,000 people to some 7,000 in the mid 1960's.

To date some 104 mega tonnes of tailings have been pro- duced by the Elliot Lake mills and are deposited as shown in Figure 2. By 1983 the ore milled will be 27,000 tonnes per day with production planned to the year 2021 for a total committed tonnage of 260 mega tonnes. There is a potential further ore reserve of 225 mega tonnes.

During the 1970's there was a resurgance in the market for uranium and in 1976 Rio Algom Ltd. decided to reactivate the Panel mine at 3,200 tonnes/day. This mine is situated on the north shore of Quirke Lake, about 20 km north of the town of Elliot Lake. It was also decided to reactivate the old Strike Lake tailings area to the north of the Panel mill. This area contained 3.26 mega tonnes from the earlier operation and could be enlarged to hold the esti- mated 11 mega tonnes of additional ore reserves.

REGIONAL PHYSIOGRAPHY

Topography and Vegetation

The topography of the Elliot Lake area is fairly characteristic of the Canadian Shield and may be described as rugged but of relatively low relief, elevation differences being generally in the order of 30 to 60 m or less. However, the entire area slopes generally upward from about elevation 275 to 300 m above sea level in the south to about 400 m above sea level in the north. In the area of the Panel tailings the elevations vary from 360 m at Quirke Lake to 390 m at Rochester Lake and an elevation of about 400 m for Strike Lake where the tailings were originally deposited. Hills near the tailings area go to a crest elevation of 460 m.

The topography is generally bedrock controlled. Topo- graphic highs consist typically of bedrock knolls or ridges and topographic lows generally contain swamps, lakes or streams. A typical view of the Elliot Lake topography is shown in Figure 3.

The area generally has a good forest cover composed of mixed coniferous and deciduous trees on the drier areas with spruce and shrubs in the damper areas. The crests of some of the higher ridges tend to have a sparse cover.

Climatology

The average daily mean temperature in the Elliot Lake area is about 4°C but varies from a daily mean of about -14°C in January to about 18°C in July (see Figure 4).

The area experiences an average total precipitation of about 79 cm (38 in.) of rainfall equivalent per year with average monthly precipitation ranging from about 5 cm per month in February to about 10 cm per month in September, October and November (see

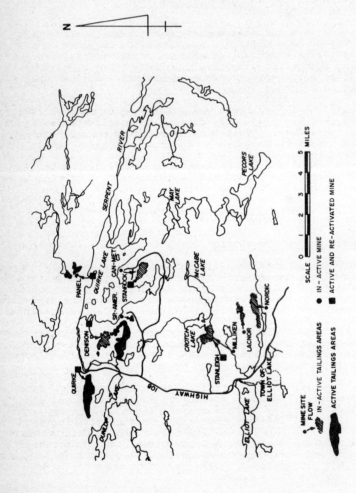

Figure 2. LOCATION OF ACTIVE AND IN-ACTIVE MINES
AND TAILINGS AREAS

Figure 3. TYPICAL ELLIOT LAKE TOPOGRAPHY

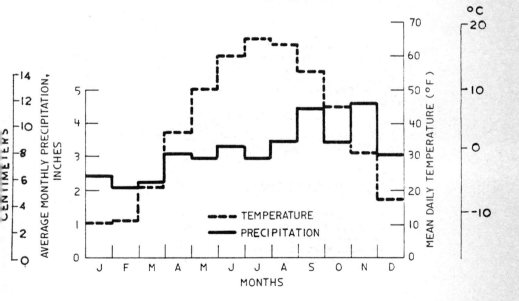

Figure 4. TEMPERATURE / RAINFALL RECORDS FOR
ELLIOT LAKE

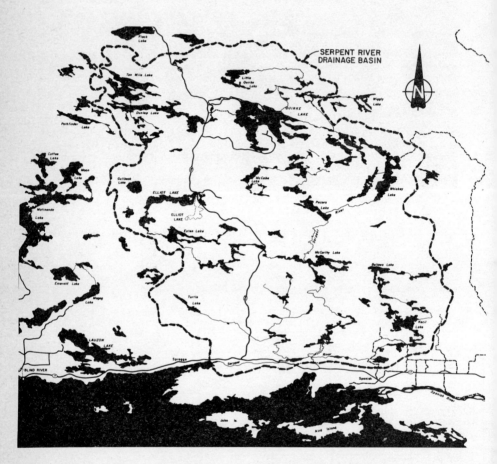

Figure 5. SERPENT RIVER DRAINAGE BASIN

Figure 4).   Of the total precipitation, about 21 cm of rainfall equivalent occurs as snow (a total of about 208 cm of snowfall) primarily during the months of November to March.

Based on 10 years of records, the maximum 24 hour precipitation in the Elliot Lake area is about 8 cm (3.2 in.). However, the design rainfall, referred to as the "regional storm" is 19 cm of precipitation in 12 hours and the maximum probable precipitation is estimated to be about 42 cm (16.7 in.) in 12 hours.

The average annual evapo-transpiration is about 50 cm (20 in.) of rainfall equivalent.   The remaining 45 cm of rainfall either infiltrates into the ground (estimated to be about 10 cm per year) or runs off directly as surface water.   Thus, as illustrated on Figures 2 and 5, the area is characterized by abundant lakes and streams;   about 20 to 25 per cent of the total area being covered by water.

Hydrology

The Elliot Lake mining area is located within the Serpent River basin which has a watershed of about 137,000 hectares (530 sq.mi.) and discharges into Lake Huron-North Channel via the Serpent River (see Figure 5).   While historic stream flow data is limited, the mean annual discharge from the basin is about $21m^3/s$ (760 cfs).

The entire basin has a uniform climate and physiography and is characteristic of watersheds in the Canadian Shield in that it experiences climatic extremes (i.e. cold winters and warm summers); a generally humid climate with uniform annual precipitation;   a thin soil cover on almost non-weathered rock;   numerous lakes, ponds and swamps;   and extensive forests.   Consequently, although there are large spring runoffs due to snow melt, the region's lakes and swamps are usually able to effectively moderate wide fluctuations in river flows resulting from short-term rainfalls.

Some 30 km to the north of the Panel mine is the White River drainage basin with its source to the north-east of Elliot Lake and draining to the south-west into Lake Huron.

REGIONAL GEOLOGY

Bedrock Geology

The bedrock in the Elliot Lake region comprises a basement complex of Archean granitic intrusives and meta-volcanics overlain unconformably by a 600 to 900 m thick sequence of metasedimentary rocks of Proterozoic age.   These rocks have been folded and faulted during the Hudsonian Orogeny into an east-west synclinal structure some 16 km wide called the Quirke syncline.   Diabase sill and dyke emplacement accompanied this period of regional deformation as did metamorphism of the sedimentary rocks to the greenschist facies. This resulted in intergranular recrystallization of the rocks, forming a very hard erosion resistant sequence which has been subdivided into five formations as illustrated on Figure 6.   Regional faults tend to occur at 5 to 6 km spacing and, with the exception of a major overthrust fault, are steeply dipping and strike generally northwest-southeast.   Local faulting tends to be more intense (0.8 to 1.2 features per square km) and, while generally steeply dipping, exhibit variable strike and persistence.

The present topography largely reflects differential weathering and erosion of the various lithological units and fault zones.

GEOLOGIC SEQUENCE

CENOZOIC
RECENT AND PLEISTOCENE
Sand, gravel and clay
GREAT UNCONFORMITY

PRECAMBRIAN
PROTEROZOIC
NIPISSING
INTRUSIVE CONTACT
11 Diabase
INTRUSIVE CONTACT
HURONIAN
COBALT GROUP
10 Lorrain Formation; Quartzite
9 Gowganda Formation; Conglomerate; Quartzite
UNCONFORMITY

BRUCE GROUP
8 Serpent Formation; Quartzite
7 Bruce and Espanola Formations; Conglomerate; Limestone, greywacke
6 Middle and Upper Mississagi Formations; Conglomerate, argillite; Quartzite
5 Lower Mississagi Formation; Quartzite, conglomerate
GREAT UNCONFORMITY

ARCHEAN
ALGOMAN GRANITE
2,4,3? Massive red granite rocks
INTRUSIVE CONTACT
3 Gneissic to massive, grey to pink, granitic rocks
RELATIONSHIP UNKNOWN
KEEWATIN (?) GROUP
1 Metavolcanics and metasediments

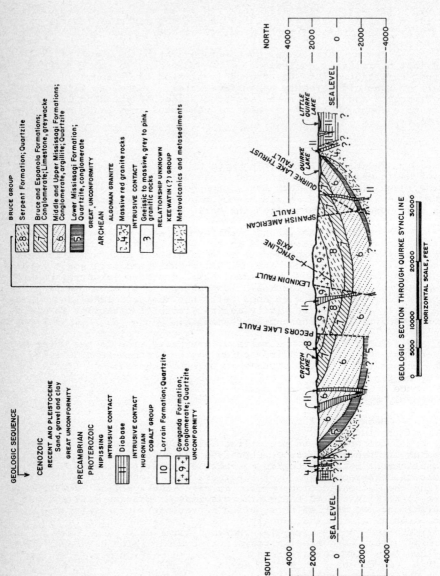

GEOLOGIC SECTION THROUGH QUIRKE SYNCLINE

HORIZONTAL SCALE, FEET

Figure 6. GEOLOGY OF QUIRKE SYNCLINE

Pleistocene Geology

While the Elliot Lake area was glaciated during each of the major continental glacial advances, the only apparent glacial deposits result from the latest, Wisconsin, period of glaciation. These deposits typically comprise well-graded silty to sandy tills laid down by the glacial ice sheet and glaciofluvial outwash deposits (sands and gravels) laid down by melt-water streams from the retreating glacier. More recently, organic accumulations have developed in poorly drained depressions or hollows. The Pleistocene deposits are generally limited in extent and are restricted to the flanks of bedrock hills or ridges (ablation tills) or the floors of bedrock valleys (fluvial sands and gravels).

Because of their limited and discontinuous occurrence, the Pleistocene deposits do not have a significant, direct impact on tailings containment and groundwater movement. However, as the only source of borrow in the area the local availability and engineering properties of the Pleistocene deposits is critical in the design of tailings dams.

Hydrogeology

As a result of metamorphic recrystallization, the intergranular porosity of the rock matrix is very low and the hydraulic conductivity of the intact rock is estimated to be of the order of 10^{-10} cm/s. Thus, for all practical engineering purposes the intact rock may be considered impervious and any potential groundwater movement through the bedrock will be controlled by joints and fractures (defects) within the otherwise intact rock. However, within areas of "unstructured" rock (i.e. rock devoid of any major structural defects such as faults), the jointing tends to be relatively widely spaced and only partially interconnected. Hence, the bulk hydraulic conductivity of the fractured rock mass tends to be relatively low and groundwater seepage through the "unstructured" rock is small. It is only in areas of intense fracturing associated with major geological features (i.e. faults and dyke intrusions) that hydraulic conductivities and hence seepage fluxes become potentially significant. However, such features are strongly oriented and localized in extent and thus the direction and point of emergence of any seepage occurring along such features can be reasonably predicted.

Because of the geological and topographical characteristics of the region, groundwater flow either within the bedrock or the overburden occurs as a complex of local, small scale, shallow groundwater systems; the concept of a "regional aquifer", consisting of a regionally extensive, permeable geological unit and associated regional groundwater flow system on the scale of tens of kilometres, is not applicable to the Elliot Lake area. Furthermore, these local flow systems are strongly influenced by the surface water systems and by groundwater mounding beneath topographic highs.

Seismicity

Based on available seismic records, it is estimated that the maximum historic horizontal ground acceleration at Elliot Lake is about 0.02 g. The results of a deterministic seismic rock analysis carried out by Weston Geophysical Research Inc. (1977) found that the Elliot Lake area is outside the four most active seismic zones in Eastern Canada and the adjacent United States. They suggest that the "Maximum Credible Earthquake" be taken as a (hypothetical) Magnitude 7.0 event located 225 km from Elliot Lake in the "Western Quebec" zone. It was predicted that this event

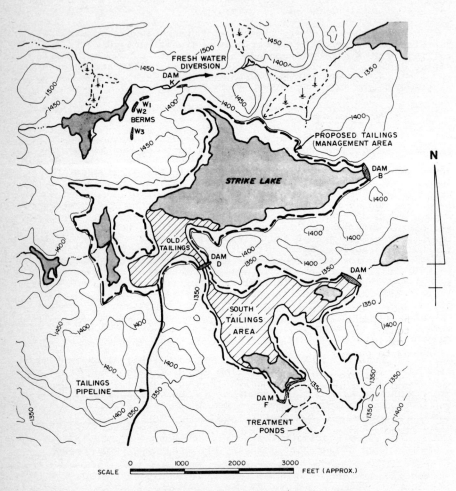

Figure 7. PANEL TAILINGS MANAGEMENT AREA

would cause an upper bound site Intensity IV (Modified Mercalli) at Elliot Lake with a corresponding peak ground acceleration of 0.065 g for average foundation conditions. The duration of shaking would be less than 20 seconds.

The seismic risk in Elliot Lake was also assessed using the probabilistic method suggested by the Division of Seismology and Geothermal Studies, Earth Physics Branch, Department of Energy, Mines and Resources, Canada. The results of this analysis indicated that for a risk level of 10^{-3} events per annum, the Elliot Lake area could be expected to experience a peak acceleration in the range of 0.018 g to 0.03 g. This risk is estimated to be associated with an earthquake ranging in magnitude from about 5.0 to 6.0 located some 50 to 100 km from the site.

THE ELLIOT LAKE ORE DEPOSIT

The uranium deposits of the Elliot Lake area occur in uraniferous quartz-pebble conglomerate associated with about 6% pyrite, near the bottom of a series of Huronian sediments lying in an east-west trending and westerly plunging syncline. At Panel, on the north limits of the syncline, the ore zone runs about 2 to 3 m thick with the dip running from 5 to 45 degrees. The production shaft intersects the ore zone at about 400 m; in the centre of the syncline the ore zone lies at about 1,500 m.

The ore may occur at one to three horizons over a vertical distance of 30 to 50 m in various areas, with the lowest bed being usually the most consistent. One characteristic of the Elliot Lake area is the general continuity of the ore over distances of several thousands of metres with few areas below economic grade. In general the area is characterized by uranium values of 0.5 to 1.5 kg/tonne.

THE PANEL MINE AND MILL

The Panel production shaft was sunk from an island close to the north shore of Quirke Lake, and mine waste rock was used to join the island to the mainland. Production was at a nominal 2,800 tonnes/day from a combination of conventional room and pillar mining to track haulage and mobile equipment haulage discharging to passes near the shaft.

The ore was crushed and ground adjacent to the production shaft and the slurry pumped to the mill that was built on the mainland about 300 m distant. Uranium extraction was by sulphuric acid leaching with the tailings being neutralized before discharge to Strike Lake area situated to the north. No tailings containment structures were constructed at this time.

When the Panel operation was reactivated in 1979 the proportion of ore mined by mobile equipment in rooms and pillars exceeded by a considerable margin the tonnage mined using conventional methods. The refurbished mill continued to use the acid leach process followed by neutralization.

PANEL TAILINGS AREA

Overall Expansion Plan

The 3.26 mega tonnes of tailings produced before 1961 had been deposited in the southwest end of Strike Lake and in a South Tailings Area south of Strike Lake (Figure 7). In order to

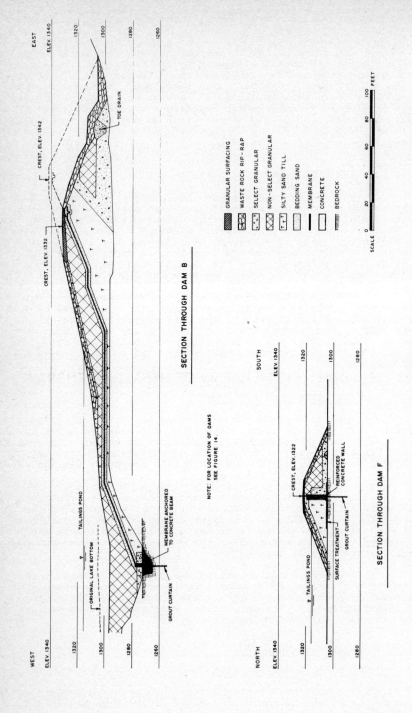

Figure 8. TYPICAL SECTIONS – DAM 'B' AND 'F'

SECTION THROUGH DAM B

SECTION THROUGH DAM F

NOTE: FOR LOCATION OF DAMS
SEE FIGURE 14.

GRANULAR SURFACING
WASTE ROCK RIP-RAP
SELECT GRANULAR
NON-SELECT GRANULAR
SILTY SAND TILL
BEDDING SAND
MEMBRANE
CONCRETE
BEDROCK

SCALE 0 20 40 60 80 100 FEET

Figure 9. INSTALLATION OF HYPALON MEMBRANE
DAM 'B'

accommodate the additional 11 mega tonnes of ore reserves, and meet
the Atomic Energy Control Board requirements current at that time
regarding seepage control etc., the following were carried out:

(a) Diverting the majority of freshwater inflow from the
     north away from the basin by the construction of Dam 'K',
     berms "W1', 'W2' and 'W3' and diversion channel 'Y'.

(b) Raising the water level in the Strike Lake basin to
     elevation 404 m to maintain a water cover over the tail-
     ings by the construction of Dam 'B' across the existing
     eastern outlet of the lake and Dam 'D' across a narrow
     valley joining the Strike Lake basin to the South Tailings
     Area.   Provision was made for subsequently raising this
     water level to elevation 407 m to accommodate future
     tailings production.

(c) Raising the water level in the South Tailings Area to
     elevation 401 m by the construction of Dams 'A' and 'F'
     to flood the existing tailings and control runoff.

(d) Decanting water from the Strike Lake basin into the South
     Tailings Area by means of a side-hill decant at Dam 'D'
     and removing effluent from the South Tailings Area via a
     decant at Dam 'F'; the treatment facility being located
     immediately downstream of Dam 'F' and discharging via
     overland flow to Quirke Lake.

     Subsequent to the construction of the dams in 1979 the
operational scheme was modified as the requirement to maintain a
water cover over the tailingswas rescinded.   As a result the
operating pond level can be maintained at about 401 m.

Containment Dams

     In accordance with governmental guidelines, all of the
perimeter dams, including Dam 'D', were to be made "impervious" and
were to be capable of storing the 100 year design storm, excess
precipitation being discharged untreated from the South Tailings
Area via an emergency spillway east of Dam 'F' (because of
operational considerations, the Strike Lake basin was capable of
storing the maximum probable flood and no interim emergency spillway was provided).

     A detailed geotechnical investigation at all proposed
dam and borrow pit sites was made.   Glacial till was available
nearby but when compacted it was unable to meet the governmental
guidelines of 10^{-6} cm/s for "impermeable" core material.   Based on
this requirement it was decided to construct the least pervious dam
practical using local materials and to also incorporate into the
design a synthetic impervious membrane.   Unreinforced hypalon was
selected because of its ductility, ease of installation and resistance to chemical deterioration.   In all cases the membrane was
secured to the rock by means of a concrete anchor beam and the rock
cement grouted to minimize seepage through joints and fissures.
Typical sections for Dams 'B' and 'F' are shown in Figure 8.   In
Figure 9 is shown the installation of the hypalon membrane at Dam 'B',
with the anchor beam already in place.

     For the operating period of the Panel Tailings Area
effluent from the Strike Lake area is discharged to the South
Tailings Area and then to a treatment plant to the south.   Here
barium chloride is added to precipitate radium and the resulting
precipitate settles out in two hypalon-lined settling ponds in
series.   The clarified effluent is discharged to Quirke Lake.

Evaluation of Tailings Containment Basin

The governmental guideline proposes a maximum average basin permeability of 10^{-5} cm/s for uranium mill tailings. In order to evaluate the average basin permeability, aerial photographic interpretation and detailed field mapping were undertaken to establish both major (i.e. faults) and minor (i.e. joints) structures around the basin. It was established that the north shore of Strike Lake was composed of diorite, the remainder of the basin being underlain by granite. The rock was more or less uniformly jointed at a 15 to 30 cm spacing with steeply dipping joints. While many of the joints showed surface apertures, borehole packer tests indicated that permeability decreased rapidly with depth.

As illustrated in Figure 10, two major faults were determined to impinge on the area of the tailings basin. The first, the Nook Lake Fault strikes northwest-southeast through the south shore of Strike Lake and to the north of the South Tailings Area. The second fault strikes northeast-southwest along the northwest shore of Strike Lake and appears to terminate at the Nook Lake Fault. In addition to these two major faults, a number of lineations were identified on aerial photos, but these were subsequently identified as being minor local faulting.

In order to determine the permeability of these major faults a number of inclined drillholes were put down to intersect the faults at about 70 m below surface. The permeabilities determined for these holes ranged from less than 10^{-6} cm/s in gouge-filled sections to about 10^{-5} cm/s.

Figure 10. PANEL TAILINGS AREA FAULTS

An evaluation of the probable average permeability value of the tailings basin indicated a value of about 5×10^{-6} cm/s. On the basis of this analysis, using normal infiltration rates, it was indicated that seepage would in general be _into_ the basin from the surrounding area except to the southeast and northeast along the faults. The total seepage along these two paths was estimated to be less than about 0.6 litres/second, which was considered acceptable. To ensure that the migration of nuclides from the tailings remains acceptable, a monitoring program has been instituted for some of the deep drillholes along the major faults.

CURRENT TAILINGS DEPOSITION SCHEME

As a result of on-going studies a deposition scheme has been developed which involves the creation of exposed tailings beaches from elevated discharge points and the construction of internal dykes to divert tailings away from the decant at Dam 'D'. With this scheme, the maximum operating pond level in the Strike Lake basin can be maintained as low as about elevation 401 m (1316 ft.) or 6 m below the previously proposed maximum level. Thus future raising of Dams 'B' and 'D' should not be necessary for current planned production.

FUTURE TAILINGS MANAGEMENT PROPOSALS

Tailings lines are being advanced along the hillside on the northwest side of Strike Lake and into the "bay" being formed to the southwest. A berm is being constructed from the old tailings in the southwest end of the lake parallel to the south shore, towards Dam 'B'.

The present interim plan is to discharge tailings from the northwest end of Strike Lake and the north and west side of "bay" to the southwest, at as high an elevation as practical. The tailings effluent from the northwest will be forced to flow around the east end of the berm being advanced towards Dam 'B' in order to reach the discharge through Dam 'D'. Effluent from the "bay" will flow through a tunnel in order to reach Dam 'D'.

The objective of the above procedure is to produce a tailings surface at close-down that has a continuous grade towards Dam 'D' and can be easily vegetated. The pond level in the Strike Lake area and in the South Tailings Area is proposed to be maintained at about 401 m (1316 ft.) elevation, with the discharge continuing to the south as at present. The above interim plan is subject to revisions based on developments in management methods up to the time of shut-down and changes in regulatory requirements.

Because the Elliot Lake area has been glaciated, surficial deposits of soil are scarce and of low value for radon retention or the growing of vegetation. It has been demonstrated that vegetation can be grown directly on pyritic uranium tailings in the Elliot Lake area, and that over the years a humus layer will accumulate. Also there are indications that within a reasonable number of years sufficient radium may have become depleted in the surface layer of tailings and transferred to greater depths in the tailings deposit so that any long term potential health hazard from radon has been greatly reduced.

Much research and investigation is in progress to arrive at a practical, economic and effective means for the long term "disposal" of uranium tailings.

THE GEOMORPHOLOGY OF THE WET AND DRY TROPICS
AND PROBLEMS ASSOCIATED WITH THE STORAGE OF
URANIUM TAILINGS IN NORTHERN AUSTRALIA

Dr. R.F. WARNER
Department of Geography, University of Sydney, Australia

Dr. G. PICKUP
Commonwealth Industrial & Scientific Research Organisation
Alice Springs, Australia

ABSTRACT

This paper describes the principal landforms of the
Alligator Rivers Region Uranium Province of Northern Australia,
reviews work on landforms and processes in this wet and dry tropical
environment, and discusses the kinds of geomorphological hazards
which might be encountered in disposing of uranium tailings at the
Nabarlek, Ranger, Koongarra and Jabiluka Uranium Project Sites.

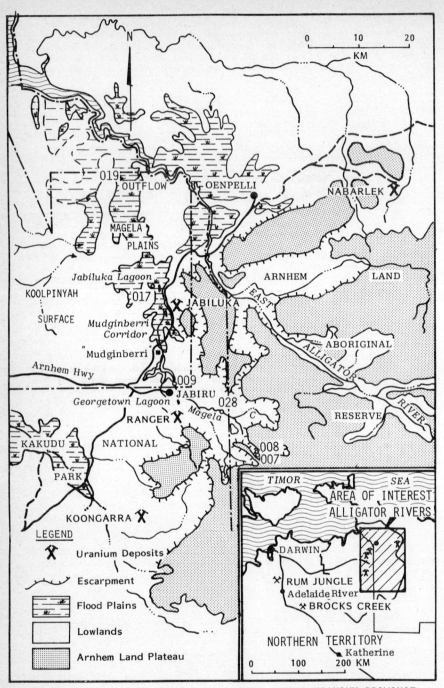

FIGURE 1 : LOCATIONS IN THE ALLIGATOR RIVERS URANIUM PROVINCE

1. Introduction

The aims of this background paper are: to describe the major landforms of the Uranium Province of Northern Australia; to review work on landforms and processes in this wet and dry tropical environment; and to discuss the kinds of geomorphological hazards which might be encountered in disposing of uranium tailings.

It begins with some consideration of the surface topography, its geology and soils. Then hydrometeorological inputs, throughputs and outflows are described briefly and finally, in the background section, modifying variables of vegetation, fauna and man are assessed. This section is followed by others dealing with: major landforms and their evolution, Pleistocene and Holocene evolution; a review of slopes, soils and related processes, and channel characteristics and fluvial erosion. The latter two are undertaken to establish natural rates of denudation. These can then be considered in the final section dealing with problems of storing tailings.

The uranium province of the Northern Territory is mainly in the catchment of the East Alligator River (14 300 km^2), although the Koongarra mine, south of Ranger, is in the South Alligator river basin (Fig. 1). The two main ore bodies are located in the catchment of Magela Creek, the major left-bank tributary of the East Alligator River. These are at Jabiluka (still undeveloped) and Ranger where ore is being mined. Mining of ore at the fourth site, Nabarlek, was completed in 1979, while at Koongarra, it is still to commence. Mining and development of these rich ore bodies was delayed by concern for the environment. A Royal Commission was set up to review much scientific evidence (Fox et al. 1977). Continuing research is being undertaken by the mining companies (Ranger Uranium Pty. Ltd. 1974 and 1975, Coffey and Partners 1980 etc.) and by the Office of the Supervising Scientist (Fry 1980), a federal agency to oversee research and regulatory procedures. Reviews of research and concern are also available from other sources (Rummery and Howes 1978, Harris 1980).

The involvement of geomorphologists has come about by the growing concern for safe rehabilitation and abandonment of mine sites, when operations have ceased. Man-made landforms will be subject to geomorphological processes and it is necessary to review both natural landform processes and those which might operate on tailings storage structures.

The geomorphological processes of this area are greatly influenced by the wet and dry tropical climate. Temperatures are always high and there is a five month wet season. The other seven months are dry. The wet season is characterised by high intensity precipitation from convectional storms, monsoons and tropical cyclones. Such rains have a very high erosion potential at the end of the dry season and in places where the vegetation cover has been destroyed or removed. In natural areas the revitalised vegetation of the wet season can retard erosional processes. However, where soils are exposed, they are vulnerable to very high rates of erosion.

The environment is incredibly rich in flora and fauna, which would react to heavy metal and radioactive pollution. The storage of mine wastes is a sensitive issue because the area is a national park of world importance. Consequently there are sound reasons for secure rehabilitation of mining sites.

2. The Physical Background

Generally land of low relief characterises the uranium province of the Northern Territory where rivers flow north into the Timor Sea. However, to the south and east, the Alligator Rivers plains are rimmed by the Arnhemland escarpment and plateau, cut in Upper Proterozoic sandstone. The plateau, which is fragmented by strong vertical jointing, is about 200 m above the plains. The latter have been cut in Lower Proterozoic metasediments, but these are now mantled by up

---

Further environmental and technical information concerning practices and proposals for the management and disposal of tailings for the Nabarlek, Ranger, Koongarra and Jabiluka Uranium Projects is provided in the Annex I of this paper.

to 30 m of Late Tertiary sediments, derived from erosion of the scarp and foothills (Story et al. 1976). Complex lateritic soils are also found on the plains, while the valley floors of this undulating lower surface are covered by up to 10 m of sandy alluvium in upstream reaches and by finer silts and clays in lower sections. In the lowest 25 km or more extensive floodplains have formed and beyond them are limited littoral deposits, both of unknown depth. Skeletal soils characterise sandstone plateaux and steeper slopes, while lateritic assemblages (Williams 1976) dominate the plains, and deep sandy soils are found on the upper alluvia: these fine to clays on the extensive lower floodplain lands (Aldrick 1976).

Climatically, the dry season is mainly from April to October and the wet, from November to March. Although a strong seasonal persistence is evident, precipitation variations within the wet season can be marked. This is reflected in variable runoff response.

Rainfall and temperature details are set out for Oenpelli (Fig. 1) in Table 1 (McAlpine 1976).

Table 1   Climatic Data for Oenpelli

	S	O	N	D	J	F	M	A	M	J	J	A	Total
M.M.P.[1]	4	28	107	226	325	302	272	68	13	2	3	1	1351
% of Total	0.3	2.1	7.9	16.7	24.1	22.4	20.1	5.0	1.0	0.1	0.2	0.1	100
M.MaT[2]	35.7	37.0	37.1	34.2	33.3	32.1	32.4	34.0	32.8	31.7	31.5	33.7	33.8
M.MiT[3]	19.6	22.1	23.4	24.0	24.2	24.2	24.1	22.9	21.1	19.0	17.8	18.1	21.7

1   Mean Monthly Precipitation for 60 years (mm)

2   Mean maximum temperature for 10 years ($^{\circ}$C)

3   Mean minimum temperature for 10 years ($^{\circ}$C)

83% of the rain falls in the four months from December to March, a period when both maximum and minimum temperatures are moderated by increased cloud cover. Highest monthly temperatures are in October and November before the rains.

Runoff responses to such a rainfall regime are fairly predictable, beginning in November with fairly rapid runoff from the shallow, stony soils of steeper slopes. Water tables rise and slowly, channels, lagoons or waterholes, and eventually flood basins, begin to fill. In the Magela flood basin (Fig. 1), outflow from the system is delayed by the great capacity of this backwater basin until January and this does not peak until March (Fig. 2). Later, higher runoff flushes lagoons, which are the only surviving surface water bodies of the dry season (Hart 1980). In this system about 28% of the rain in the upper catchment runs off (Fox et al. 1977). Percentages of runoff into the flood basin may increase in the lower parts catchment, but much of this appears to evaporate rather than flow out into the East Alligator River (N.T.A.D.T.W. 1980).

Figures for 1978/79 reveal the nature of runoff in the Magela catchment (Fig. 2). In that year rainfall was above average (1500 mm). Table 2 shows inflow and outflow data for the major part of the backwater basin. The gauging station at Jabiluka is near the top of the basin and that at the Outflow is the lowest station above the junction with the East Alligator River (Fig. 1).

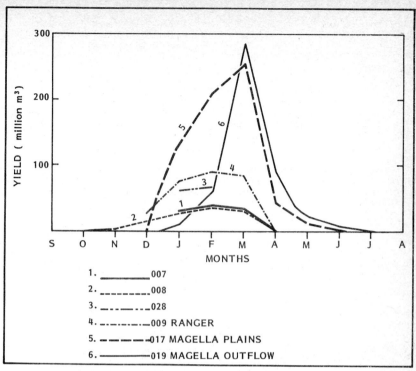

FIGURE 2 : MONTHLY DISCHARGES FOR MAGELA CREEK
GAUGING STATIONS 1978/79.

Table 2   Magela Runoff at Jabiluka and Outflow, N.T.
(based on N.T.A.D.T.W. 1980 data)

	Jabiluka		Outflow	
	mill. m^3	cumulative	mill. m^3	cumulative
Dec.	2.7	2.7	0	0
Jan.	132.0	134.7	11.5	11.5
Feb.	207.6	342.3	59.3	70.8
Mar.	254.4	596.7	283.9	354.7
Apr.	44.3	641.0	92.0	446.7
May	12.1	653.1	26.0	472.7
Jun.	3.1	656.2	7.8	480.5
Jul.	0	656.2	2.4	482.9

In the first three months, inflow is greater than outflow, and although outflow was higher from March to July (5 months), it was still 173.3 mill.m^3 less than the inflow and yet the former includes a further 431 km^2 of catchment area.

Table 3 reveals that losses to ground water and evaporation/transpiration may be about 60% in the upper two-thirds of catchment. These losses increase, mainly in evaporation from the basin, to nearly 80% for the whole basin.

Table 3   Data for Jabiluka and Outflow

	Inflow (Jabiluka)	Between	Outflow
Area (km^2)	1134	431	1565
Precip. (mill. m^3)	1701	647	2348
R.O. (mill. m^3)	656	-173	483
Evap. + Grd. water (mill. m^3)	1045	820	1865
R.O.%	39%		21%

On the plains, savannah woodlands and grasses predominate, while in the wetlands, a wide range of flora, from sedges to paper bark forests, is found (Story et al. 1976). Vegetation plays a vital role in the geomorphology of this landscape (Story et al. 1976, Williams 1976, Fox et al. 1977 and Richards 1978). In the wet season it protects the surface from rainsplash and helps to lessen the impacts of high intensity precipitation.

The fauna also play an important part in the geomorphology of this area. Water buffaloes were introduced in 1826 and now about 150 000 run wild, having a marked effect on grasses, puddling soils around waterholes and in creating wallows (Story et al. 1976). Termites have modified denudation rates and soil-forming processes in some areas (Williams 1968, 1976 and 1978), while worms have influenced soils in piping for water entry and movement down to weathering fronts.

Land clearance for roads, tracks and for mining, have been important in accelerating natural erosion rates, as has the effects of overgrazing. The difference between natural and man-induced rates are a major concern in assessing likely rates of denudation in abandoned mine workings.

3. Major Landforms and their Evolution

The major elements of the landscape, the plateau and plains, are old features; the former dating from a post-Lower Cretaceous planation (Wright 1963, Hays 1967, Galloway 1976 and Williams 1976). This planation has been referred to by several names, more recently the Bradshaw surface (Williams 1969).

This surface was truncated early in the Tertiary by a new and lower surface cut in the less resistant metasediments, where slopes now are generally less than 2°. The composite origins of the plains are reflected in the thin Tertiary sediment mantle and in the lateritic soils. This planation is known as the Koolpinyah surface (Williams 1969) and is regarded as Late Tertiary in age.

The flatter slopes are flanked by a variety of marginal slope forms, characterised by a range of lateritic materials. Williams (1976) estimated that 40% of Koolpinyah surface was pisolitic blocks, being destroyed by current erosion and weathering, 20% was in mottled outcrops with gravel veneering, indicating prior truncation, and another 20% was composed of disintegrated fragments.

4. Pleistocene and Holocene Evolution

Pleistocene and Holocene climatic changes and variations in sealevel have modified major forms through rejuvenation and sedimentation in lower drainage reaches, through changed weathering and denudation rates. The evidence is seen in changed soil profiles, in the degradation and aggradation of river channels and flanking alluvia. Soils are now polygenetic or complex (Williams 1976, Story et al. 1976). River profiles have been incised and are flanked in parts by ancient flood plains or terraces, and the last major transgression produced the aggradation forms now dominant in lower river courses. The more active deposition in the lower East Alligator River lead to the creation of the unique wetlands flood basin in the lower part of Magela Creek. This major tributary of the East Alligator drains mainly plainlands and, although sandy braided channels exist in the upper parts, the lower parts are dominated by finer sediments and the rate of sedimentation has been less than in the main river. So in spite of the fact that pronounced levees along the East Alligator River have not been found in recent surveys (N.T.A.D.T.W. 1980), an extensive backwater area does exist. High flows in the main river cause backup and ponding of Magela flows.

5. Slopes, Soils and Processes

Under natural conditions, the great variety of soils (Hooper 1969 and Aldrick 1976) are denuded at rates dependent on the intensity, duration and timing of rainfall, the nature of the vegetation cover and the physical character of the slopes. The environmental concern about the integrity of man-made landforms and their likely denudation, necessitates some appraisal of the range of local rates of creep and slopewash. Experiments carried out by Williams (1969, 1973, 1976 and 1978) south of Darwin on gratitic and sandstone slopes are useful.

He was concerned with soil creep, the slow downhill movement of debris and soil under gravity, and slopewash, the detachment and removal of soil particles by raindrop impacts, by overland flow and by the possible addition of seepage flows on foot slopes. Creep was measured using "Young pits" with iron rods buried at 10 cm intervals. Wash materials were caught in trays with rims at soil-surface level. The granitic soils were either deep colluvial profiles or three-layered soils, ranging from a sandy surface, through quartz fragments, to weathered granite. Mound-building termites are thought to be responsible for the latter profiles (Williams 1969 and 1978, Lee and Wood 1971). Deep soils or sandstones are rare on slopes above 3° and soils have involved a range from texture-contrast profiles on alluvium to rocky lithosols on slopes (Williams 1969 and 1973).

Mean creep rates on the granites and sandstones were 7.33 and 4.39 cm^3/cm/yr respectively, while mean slopewash rates were 53.6 and 55.8 mm/1000 yr. The overland flow erosion on sandstones was higher at the base of 5° footslopes (1830 cm^3/yr) than on 11° hillslopes (680 cm^3/yr) (Williams 1976). This was attributed to: (a) permeability of upper slopes, (b) protection of mantles by stones, (c) stronger texture contrasts in colluvial and alluvial soils which are saturated more and have less cohesion, and (d) greater importance of seepage on lower slopes. Wash rates in the granites were five times higher than creep and on the sandstones 5.2 times higher. This is attributable in part to the sparse vegetation at times. Rock creep of individual particles ranged from 1 cm/yr on 3° to 3 cm/yr on 18° slopes (Williams 1976).

Richards (1978) distinguished between natural or geological erosion and man-induced erosion, where accelerated processes may be increased up to 1000 times the natural rate. The geological rate is essentially at the rate of soil forming or weathering, because most soils on slopes over 3° are skeletal. He claimed that up to half the annual soil loss could occur in one or two events, where rainfall intensities of 30 mm/20 mins. may be attained each year. Erosion rates can be 20 to 40 times greater at the start of the wet season (Williams 1976), while a 95% reduction in these rates can be effected by 40% ground cover development (Richards 1978).

Bare ground is very susceptible to erosion in the mid-wet period by the high intensity rains and erodibility is then a function of the physical character of the soils and their management (Richards 1978). Non-cohesive sandy soils are readily damaged following clearance, overgrazing and traversing by off-road vehicles. Gullies develop rapidly in wheel ruts, on cattle pads and in fire breaks, with increases in slope angle and lengths adding to the risks.

Thus there are five major processes operating outside the channels (Williams 1976):

(a) Mass movement, which includes soil creep (slow compared with slope wash or overland flow erosion), surface rock creep which increases in a linear manner with increased slope angle and landslides (not common, although rock falls do occur on the escarpment).

(b) Rain splash erosion, a major mechanism for soil removal by high intensity rainfall, particularly at the start of the wet season, when the rate may be 20 to 40 times higher than later in the wet season.

(c) Overland flow, essentially a function of slope length, gradient, soil type, precipitation intensity and duration, and surface cover. This is much higher in footslope areas than on steep, rocky hillslopes. It is up to five times greater than creep.

(d) Throughflow, more active in the wet season, when profiles have been deeply infiltrated by water and when downslope movements of water do appear to increase rates of footslope erosion.

(e) Gully erosion, channel formation by runoff concentration in tracks made by animals and vehicles, and representing a much accelerated form of denudation.

Williams (1976) cited three examples of rapid erosion:

(a) A 17.5 cm deep gully, formed from an area of 800 m^2 at Jabiru, yielded 100 m^3 of coarse sand in two years. This is a denudation rate of 62,500 m^3/km^2/yr or 125 mm/yr, 250 times the predicted rate for rehabilitated tailings surface soils after abandonment (Costello, pers. comm.).

(b) Road cuttings left at above 11° are prone to slumps and slides as seen on the Stuart Highway.

(c) At Brock's Creek 200 km south of Darwin in 1966/67 a fall of 135 mm from a cyclone yielded 11 m^3/km^2 (about 1/8th the annual total) in two days. Thus the incidence of cyclones and rainsplash from more than 100 mm/day of rain have to be carefully considered in land clearance problems.

6. Channel Characteristics and Fluvial Erosion

The major landforms and geology influence the channel type, long-profile changes, and with precipitation inputs, influence flow patterns and volumes, together with sediment loads. Thus, taking the Magela system, several types of channel may be defined:

(a) Upper bedload tracts of limited extent on the Arnhemland plateau. There bedrock sandstone conditions prevail and channels are often straight, influenced by strong vertical jointing.

(b) Upper gorge tracts, where the rivers have cut back into the escarpment. The channels here are flanked by high sandstone bluffs. The break between upper and lower levels is often marked by a waterfall. Again, bedload channels prevail with no great development of floodplain.

(c) Upper braided tracts. When the channels emerge from the gorges the valleys are much more open, with a sandy bedload extending into the plain area. Two or three channels exist separated by tree-covered braids, indicating lateral channel movement is not rapid. Gradients are still steep enough to allow velocities of 0.3-0.4 m/sec at 20 m^3/sec. Between a near-gorge station (028) and Ranger (009), the mean slope is 0.0007. Waterholes are often located in lower tributaries, such as at Georgetown, where sandy levees flanking the braids partly dam tributary waters. These are the backflow billabongs of Hart (1980), so called because high flows in the main stream can flow back into the tributaries.

(d) The braided channel then gives way to what is called the Mudginberri Corridor, a series of shallow stream channels, which dry out in the dry season, billabongs and floodplains (Hart 1980). These channel billabongs are sandy, deep (6 m) and clear; they are well flushed by the main flows and are probably fed during the dry season by seepage from adjacent sandy aquifers. Only the waterholes survive the seasonal drought and the poorly defined, shallow channels dry up. These areas are dominated by paper bark forests. Here gradients are much flatter. Sandy deltas mark the head of the main lagoon and velocities are much less, as the water spreads out through a broader confused drainage pattern (0.04-0.35 m^3/sec) (N.T.A.D.T.W. 1980).

(e) Below Jabiluka Lagoon, marking the end of the corridor, there is the so called Magela floodplain (Fox et al. 1977) (155 km^2). This is not a floodplain in the normal geomorphological sense, subject to temporary inundation following surcharging by the annual flood ((Leopold, Wolman and Miller 1964). It is an area up to 200 km^2 in normal years which is inundated for the later half of the wet season. Much of the water appears to be lost to evaporation from an area which includes paper bark forests, sedgelands, swamps and so on. Once in ten years it may extend to over 300 km^2 and up to 500 km^2, once in 100 years (Fox et al. 1977). It is essentially a backwater flood basin related to the discordant junction with the East Alligator River, and has a very low gradient (0.000004).

The geomorphological characteristics of the Magela catchment and its distinctive regime are important because controlled or accidental releases of polluted waters from Jabiluka or Ranger could in the long term impair the quality of the wetlands environment.

This pattern of long-profile distinctiveness in the tropics is not unusual because similar profiles have been described for the Fly and Purari Rivers in Papua New Guinea (Pickup in prep., Pickup and Warner in prep.). The major difference in this case is the absence of armoured reaches caused by gravel and boulder bedloads. Long-profile and cross section details to demonstrate the nature of these adjustments are not available, except for recent cross-section and long-profile surveys below Mudginberri (N.T.A.D.T.W. 1980). 35 cross-sections from near Ranger to the Outflow have been completed, but details of 28 given are mainly for the corridor and flood plain.

Data on rates of denudation for this system are limited to a few suspended sediment concentration samples in 1978/79. These are generally low (5-12 mg/l for individual flow hydrographs), but concentrations of up to 82 mg/l have been recorded for runoff from Corndorl Creek, draining part of Jabiru township (N.T.A.D.T.W. 1980). More observations, as yet unpublished, will improve knowledge of local denudation.

Fluvial erosion is obviously highly seasonal, with probably high rates prevailing in the first major storm of the wet season, prior to reestablishment of vegetation. However, limited data, reported by Williams (1976), for the Adelaide River which is a similar type of system, suggest that much of the erosion occurs in March (4800 t or 70%) and February (2000 t, 29%). The total yield for this catchment is only just over 6800 t/yr or 6.9 m^3/km^2/yr. This is an annual rate of lowering of about 0.0069 mm/yr. Accelerated erosion for limited areas could increase this by 100 to 1000 times (0.7-7.0 mm/yr). The Ranger example cited earlier involved 62.5 mm/yr, nearly 10,000 times more than the natural rate. Thus until more data are available, a crude estimate of natural denudation would be about 0.007 mm/yr but this can obviously be increased by several orders of magnitude on local disturbed surfaces. Thus most of the natural catchment has a fairly low denudation rate. Where high rates do occur, they are generally for small areas and their total impact on the system is quite small. However, high rates in polluted materials could have far-reaching effects in local water-based ecosystems.

In Magela Creek, the load derived from the upper catchment is mainly deposited in the backwater basin. The figures quoted above indicate something like 7,000-10,000 m^3/yr of sedimentation or about 0.05 mm/yr on 200 km^2 of floodplain (or about 0.25 m in the last 5,000 years). Such a figure seems far too low at this stage and the concentrations measured elsewhere may be ignoring a large part of the total loads. However, estuarine deposits have been found under shallow alluvium in lower parts of the flood basin (Galloway 1976).

7. Problems of Storing Tailings

Having discussed the geomorphology of the uranium province largely in terms of natural denudation, it is now necessary to review some of the geomorphological problems which may confront the managers of mine sites, particularly those being rehabilitated prior to abandonment.

The impacts of removing vegetation in this sensitive environment have been briefly discussed through examples of very high rates of erosion. Such knowledge must be used, together with other data when it becomes available, in trying to vegetate and stabilise man-made landforms such as tailings dumps. The example of Rum Jungle, a uranium mine abandoned some 20 years ago, has often been cited as an example of the kind of pollution which may follow fluvial breaching of tailings dumps (Davy 1975 and 1978).

The technical aspects of abandoning disposal sites are just part of the total problem. Another part must include the assessment of the hydrometeorological impacts and the land-based responses to these events which may occur in the future. Time scales involved can involve thousands of years, as was outlined by Rafferty (1980).

At this stage, some preliminary discussion of the hazards which can further accelerate denudation after site rehabilitation needs to be undertaken. The rupture or destruction of tailings storage capping could cause major problems if large slugs of tailings were accidentally added to slope and fluvial systems. The Ranger tailings dam will contain 27-45 mill. m^3 of waste material (Fox et al. 1977).

Climatic hazards undoubtedly provide the water and wind which could erode or deflate protective caps and marginal slopes of the structures. Major tropical storms have marked effects on the natural surfaces, particularly if received before vegetation recovery from the seasonal drought. When such falls occur in areas devoid of vegetation, they have even greater impact later in the wet season because of higher surface impact momenta and intensity. 10, 100 and 1000 year estimates of annual precipitation have been worked out for this area (S.M.E.C. 1975) mainly for evaporation purposes but these are of little value in predicting denudation. The once in 1000 year total is less than twice the average annual total. Storms of over 100 mm/day are common, with probably one or more each year. The following figures are from Australian Rainfall and Runoff for this area:

2 yr 12 hr intensity 7-8 mm/hr (80-100 mm/12hr)
50 yr 12 hr intensity 16 mm/hr (192 mm/12 hr)
2 yr 72 hr intensity 2.4 mm/hr (173 mm/72 hr)
50 yr 72 hr intensity 5.5 mm/hr (396 mm/72 hr)
1 yr-100 yr 1 hr intensity 50-100 mm/hr
1 yr-100 yr 24 hr intensity 4-11 mm/hr (Aust. Inst. Eng. 1977)

With events of such intensity, magnitudes and frequencies, the state of the land surface becomes vitally important in determining erosional response. Thus it becomes necessary to consider the soil or fill type; slope and length of slope; vegetation type and its ability to survive, as well as to bind the regolith; the shape or topography of the forms; ground-water conditions and so on.

An impervious soil or fill, even on a near flat surface, will result in a high percentage of runoff and this will detach soil particles, even under favourable vegetation conditions, particularly towards the down-slope margins of structures, especially where relatively steep. The natural soils of this region are often stony and pervious, with stonelines acting like an armouring mechanism (Williams 1973 and 1976). A pervious fill would admit much water which would then be able to react with pollutants. These could then seep into groundwater or cause a perched water table within the storage. The former could pollute vegetation and water beyond the structure, while the latter could, if sufficient elevation were reached, affect the root systems of vegetation growing on the capping materials. Thus it seems imperative to admit as little water as possible to the main tailings and yet sufficient upper infiltration and storage to sustain cap vegetation. Seasonal precipitation failure during the late wet season plus fire could make it difficult to maintain an effective vegetation cover. Colonising soil piles in which sulphide pollutants were present would not be easy (Davy 1978), although it is thought to be possible (Morley 1978).

Slope conditions are obviously important because it is necessary to have some drainage. Slopes of about 11° are too steep, unless stone faced, because slumping would occur (Williams 1976). Any slope above 2° is subject to some creep (Williams 1974), especially under poor compaction. Therefore low angles, good compaction and/or facing with stone, bitumen or other artificial materials are important to retard marginal rilling, gullying, slope wash and creep.

Whilst experiments using various types of vegetation have been foreshadowed (Fox et al. 1977 and Morley 1978), it will need to be grown on exposed surfaces, where soil or fill type, depth, water supply and pollutants will have to be considered. The introduced vegetation needs to survive the long seasonal droughts and regenerate rapidly enough in the wet season to have a desirable soil-holding capacity. This will need detailed research, especially for the conditions in this environment. Studies on coal dumps in South Wales by Haigh (1979 and 1980) have revealed fairly high rates of surface lowering under both well- and poorly-vegetated conditions. On vegetated main slopes and upper convexities, the rate was 1.9 mm/yr and on less vegetated profiles 6.2 mm/yr., but these were slopes of 18°. Lusby and Toy (1976) have also used rainfall simulation to evaluate restoration on mine spoils in Wyoming. They found that rehabilitated slopes erode more rapidly than natural slopes at rates of two to four times greater in vegetated areas and at much higher rates where vegetation cover was poor. Soil yield was seen as a good parameter because it integrated slope effects, soil and vegetation properties.

The topography of abandoned dumps should blend aesthetically with the natural landscape (Morley 1978) but contouring and shaping are important to help evacuate water, to alleviate ponding and soil saturation. High evaporation from dissolved mineral-rich fills could introduce upper profile salting.

Tailings dams have been planned and designed with considerable care (Ranger Mines Pty. Ltd. 1974 and 1975, Coffey and Partners 1980, etc.) to minimise seepage through the wall and from the base into ground water. Trenches are used to monitor seepage and leaked materials may even be pumped back into active tailings ponds. When abandoned and with the water cover removed, such seepage may be considerably less. It is however still important to minimise this loss because contamination of ground water and waterholes in the dry season would represent major ecological disruptions. Waterholes and billabongs are important survival habitats in the seasonal drought.

Monitoring and modelling of slope and fluvial processes will provide more detail to help make ultimate decisions about tailings storage. This kind of research is already being pursued in the hydrology and dispersion studies of Magela Creek (Smith, Young and Goldberg 1978, Young 1980a and 1980b). Like geomorphology, even this work is dependent not only on process data, but also on dimensional information, particularly where morphologies are unstable.

8. Conclusions

This paper set out to provide background to the geomorphology of the wet and dry tropics of the Northern Australian Uranium Province. The present geomorphological processes are strongly related to the seasonal high intensity summer rain and to the effectiveness of the natural vegetation cover. Under natural conditions it is apparent that rates of denudation do not differ greatly from some temperate humid areas. However, it is very clear that when the balance maintained by vegetation is disrupted or destroyed, rates of denudation may be accelerated by up to 10,000 times. This poses great problems in mine and environmental management, which will be particularly severe with abandonment of tailings dams. Active research is proceeding in many fields already, but this must include geomorpholigical contributions, especially if man-made landforms are to retain their integrity for hundreds of years.

9. *Summary of the Workshop*

A summary of geomorphological considerations associated with the siting of uranium mill tailings disposal dams and their abandonment after decommissioning and rehabilitation, which was written by Dr. Warner following the Workshop, is provided in Annex II.

10. Acknowledgements

The authors acknowledge discussions with staff members from the Australian Atomic Energy Commission and from the Office of the Supervising Scientist.

11. References

Aldrick, J.M. (1976) Soils of the Alligator River area. in R. Story et al.. Lands of the Alligator Rivers Area, Northern Territory. C.S.I.R.O. Land Research Series, No. 38, 71-88.

Australian Institution of Engineers (1977) Australian rainfall and runoff : flood analysis and design.

Coffey and Partners (1980) Construction foundations of the tailings dam and water retention embankments. Report S6274.

Davy, D.R. (ed.) (1975) Rum Jungle Environmental Studies. Aust. Atom. Energy Comm. Report E.65.

Davy, D.R. (1978) Uranium – Overview of the Alligator Rivers area, Northern Territory, Australia. in R.A. Rummery and K.M.W. Howes (eds.) Management of Lands Affected by Mining, C.S.I.R.O. Div. of Land Resources Management Workshop, Kalgoorlie. 71-86.

Fox, R.W. (1977) Ranger Uranium Environmental Inquiry. 2nd Report Aust. Govt. Publisher, Canberra.

Fry, R.M. (1980) Environmental protection and uranium mining in the Alligator Rivers Region. in S. Harris (ed.) Social and Environmental Choice : the Impact of Mining in the Northern Territory. CRES Monograph No. 3, Aust. Nat. Univ. 12-24.

Galloway, R.W. (1976) Geomorphology of the Alligator Rivers area. in R. Story et al. Lands of the Alligator Rivers Area, Northern Territory. C.S.I.R.O. Land Research Series, No. 38, 52-70.

Haigh, M.J. (1979) Ground retreat and slope evolution on regraded surface mine dump, Wannafon, Gwent. Earth Surface Processes, 4. 183-189.

Haigh, M.J. (1980) Slope retreat and gullying on revegetated surface mine dump, Waun Hoscyn, Gwent. Earth Surface Processes, 5. 77-80.

Harris, S. (ed.) (1980) Social and Environmental Choice : the Impact of Mining in the Northern Territory. CRES Monograph No. 3, Aust. Nat. Univ.

Hart, B. (1980) Water quality and aquatic biota in Magela Creek. in S. Harris, (ed.) Social and Environmental Choice : the Impact of Mining in the Northern Territory. CRES Monograph No. 3, Aust. Nat. Univ. 47-63.

Hays, J. (1967) Land surfaces and laterites in the north of the Northern Territory. in J.N. Jennings and J.A. Mabbutt (eds.) Landform Studies from Australia and New Guinea. A.N.U. Press, Canberra. 182-210.

Hooper, A.D.L. (1969) Soils of the Adelaide-Alligator area. in R. Story et al. Lands of the Adelaide-Alligator area, Northern Territory. C.S.I.R.O. Land Research Series, No. 25. 95-113.

Lee, K.E. and Wood, T.G. (1971) Physical and chemical effects on soils of some Australian termites and their pedological significance. Pedobiologia, 11. 376-409.

Leopold, L.M., Wolman, M.G. and Miller, J.P. (1964)   Fluvial Processes in Geomorphology. Freeman, San Francisco.

Lusby, G.C. and Toy, T.J. (1976)   An evaluation of surface - mine spoils area restoration in Wyoming using rainfall simulation. Earth Surface Processes, 4. 375-386.

McAlpine, J.R. (1976) Climate and water balance. in R. Story et al. Lands of the Alligator Rivers Area, Northern Territory. C.S.I.R.O. Land Research Series, No. 38.

Morley, A.W. (1978) Rehabilitation in vein deposit mining - the Jabiluka uranium project. in R.A. Rummery and K.M.W. Howes (eds.) Management of Lands Affected by Mining. C.S.I.R.O. Div. of Land Resources Management Workshop, Kalgoorlie. 87-97.

N.T.A.D.T.W. (1980) Uranium Province hydrology Vol. 11. Report by Hydrology Section, Water Division, Northern Territory Department of Transport and Works.

Pickup, G. (in preparation) Geomorphology of tropical rivers. II Adjustment and Equilibrium.

Pickup, G. and Warner, R.F. (in preparation) Geomorphology of tropical rivers. I geomorphic history, hydrology and sediment transport in the Fly and Purari, Papua New Guinea.

Rafferty, P.J. (1980) Application of geomorphology to engineering of uranium mill tailings disposal sites for long-term stability - working paper. OECD - Nuclear Energy Agency, Paris.

Ranger Uranium Mines Pty. Ltd. (1974 and 1975) Environmental impact statement and supplements 1 and 2.

Rummery, R.A. and Howes, W.M.W. (eds.) (1978) Management of Lands Affected by Mining. C.S.I.R.O. Div. of Land Resources Management Workshop, Kalgoorlie.

Richards, R.J. (1978) Soil erosion hazard in the Northern Territory. in G. Pickup (ed.) Natural Hazards Management in North Australia. Proceeding and papers from 2nd NARU Seminar, Darwin, N.T. 247-281.

S.M.E.C. (1975)   Ranger uranium mines tailings retention system: design assessment. Snowy Mountains Electricity Commission Report for Aust. Atomic Energy Commission.

Smith, D.I., Young, P.C. and Goldberg, R.J. (1978) Provisional summary report of Magela Creek dye tracing experiments. CRES Applied Systems Program Working Paper AS/WP5.

Story, R., Williams, M.A.J., Hooper, A.D.L., O'Ferrall, R.E. and McAlpine, J.R. (1969) Lands of the Adelaide-Alligator area, Northern Territory. C.S.I.R.O. Land Research Series, No. 25.

Story, R., Galloway, R.W., McAlpine, J.R., Aldrick, J.M. and Williams M.A.J. (1976) Lands of the Alligator Rivers area, Northern Territory. C.S.I.R.O. Land Research Series, No. 38.

Williams, M.A.J. (1969) Geomorphology of the Adelaide-Alligator area. in R. Story et al. Lands of the Adelaide-Alligator Area, Northern Territory. C.S.I.R.O. Land Research Series, No. 25.

Williams, M.A.J. (1973) The efficiency of creep and slopewash in tropical and temperate Australia. Aust. Geog. Studies., 11. 62-78.

Williams, M.A.J. (1974) Surface rock creep on slopes in the Northern Territory of Australia. Aust. Geogr., 12. 419-424.

Williams, M.A.J. (1976) Erosion in the Alligator Rivers area. in R. Story et al. Lands of the Alligator Rivers Area, Northern Territory. C.S.R.I.O. Land Research Series, No. 38. 112-125.

Williams, M.A.J. (1978) Termites, soils and landscape equilibrium in the Northern Territory of Australia. in J.L. Davies and M.A.J. Williams (eds.) Landform Evolution in Australia. A.N.U. Press. Canberra 128-141.

Wright, R.L. (1963) Deep weathering and erosion surfaces in the Daly River Basin, Northern Territory. J. Geol. Soc. Aust., 10. 151-164.

Young, P. (1980 a) Mining and natural environment : water and the dispersion of pollutants. in S. Harris (ed.) Social and Environmental Choice : the Impact of Uranium Mining in the Northern Territory. CRES Monograph No. 3, Aust. Nat. Univ. 25-46.

Young, P. (1980 b) Mining and the natural environment : systems analysis and mathematical modelling. in S. Harris (ed.) Social and Environmental Choice : the Impact of Uranium Mining in the Northern Territory. CRES Monograph No. 3, Aust. Nat. Univ. 64-78.

FURTHER ENVIRONMENTAL AND TECHNICAL INFORMATION CONCERNING
PRACTICES AND PROPOSALS FOR THE MANAGEMENT AND DISPOSAL OF TAILINGS
FOR THE NABARLEK, RANGER, KOONGARRA AND JABILUKA URANIUM PROJECTS

A summary of details presented at the Workshop by
R.F. WARNER, P.J. BURGESS and D.R. DAVY

Four major ore bodies have been found in the Alligator Rivers Uranium
Province of the Northern Territory, Australia, which is located about
200 km east of Darwin. These are at Ranger, Jabiluka, Koongarra and
Nabarlek (Fig. 1). Of these Nabarlek has completed mining operations
and the ore is being milled over a nine-year period. Ranger has
commenced mining the No.1 of the five ore bodies discovered and operations
are expected to last for about 25 years. Jabiluka and Koongarra have yet
to start mining but planning and most impact assessments are complete.

All four sites have ore bodies in the Koolpin Formation, a Lower
Proterozoic metasediment complex. All mine sites too are formed on a
fairly low, deeply weathered, lateritic plain, known as the Koolpinyah
surface (Table 4). This is a Late Tertiary erosion surface located
below the sandstone residuals of the Arnhemland Plateau (Figure 1).
In each case the siting of the tailings dams has involved (or will involve)
minimal disturbance of a very sensitive natural surface. In the two
examples presented, Ranger and Koongarra, the tailings dams are sited
at the heads of minor drainage systems. This way impacts on catchments
are minimised.

The area is characterised by a hot, wet and dry, tropical climate.
Temperatures fluctuate between about 18 and 38°C, with diurnal being
greater than annual variations. The precipitation of about 1300-1500
mm is mainly received in the five months from November to March in high

Table 4 :   Summary of Site Details for Uranium Mines in Alligator Rivers

	RANGER	NABARLEK	JABILUKA	KOONGARRA
Topography	Small catchments W of Magela Ck - on Koolpinyah Surface	Small catchment W of Cooper Ck, below sandstone outcrops on Koolpinyah Surface	Edge of Magela Floodplain in Mudginberri Corridor below sandstones on Koolpinyah Surface	Small catchments at the head of Nourlangie Ck, below Mt.Brockman sandstones, on Koolpinyah Surface
Drainage	Small streams incised into Koolpinyah Surface	Small incised streams	Adjacent to Magela flood plain	Small incised streams on the Koolpinyah Surface
Ore Reserves (tonnes)	124,700	12,000	206,000	15,500
Waste Rock/ Ore Ratio	3.1	8.2	?	-
Tailings (tonnes)	55×10^6	4.9×10^5	55.3×10^6	2.2×10^6
Waste Rock (tonnes)	135×10^6	4×10^6	10.3×10^6	8.5×10^6
Tailings Dam Crest Height (m)	30	?	38	12
Area (ha)	110	13	192	2×14
Annual Seepage (m)	3.2×10^5	3.6×10^4	4.3×10^5	1.8×10^4
Disturbed Area Tailings Waste Rock Total	123* 150 680	13 10 121	210* 292 1122	28 33 225
Disposal of Tailings	At present above grade but should be returned to mine pit	Subgrade mine pit	Above grade (also in underground workings)	Below grade in specially excavated pits

* Ultimately nearly twice as large

(Sources: Data are subject to revision.   Compiled mainly from:
 D.R. Davy: An Overview of the Alligator Rivers Area, Northern Territory,
   Australia, in R.A. Rummery and K.M.W. Howes (eds.) Management of Lands Affected
   by Mining, C.S.I.R.O. Div. of Land Res. Management, 1978, 71-86.

Various environmental impact statements and consultants' reports.)

intensity convectional, monsoonal and cyclonic storms. In the early
part of this period, the vegetation is able to recover from the seven-
month seasonal drought and it is able to restrict natural denudation to
fairly low rates. However, out-of-season or early downpours on dried up
or burned vegetation can cause rapid erosion. Wet-season impacts are
also high on disturbed surfaces. Evaporation of about 2200 mm exceeds
precipitation in most years, but in wet years when evaporation is lower,
there can be problems in managing water at the mines.

The vegetation is mainly grasses and Savannah woodland, but large parts
of the lower drainage areas are characterised by wetlands. The notable
example is the Magela floodplain, below Ranger and adjacent to Jabiluka
(Fig. 1), which is a 200 km² seasonally inundated backwater system.
This is dominated by sedge and herbaceous swamps, and by paperbark
forests.

The Ranger tailings dam is located at the top of Coonjimba Creek, a
west-bank tributary of Magela Creek. The containment area is about
110 ha. The walls of this above-grade structure are elevated as required,
to allow the progressive buildup of tailings, always to be stored below
water. Eventually the crest height could reach 30 m or more, with a
capacity of between 28 and 55 mill m³ (Table 4).

Seepage from this facility is restricted by a cut off in the base of the
dam reaching low permeability in weathered rock and by grouting of more
permeable zones.

The floor of the impoundment is in weathered, clay-rich rocks.
Ultimate rehabilitation plans call for the disposal of tailings into
the mine pit, although other alternatives are being considered.
These may involve the removal of the water in the tailings dam and the
capping of waste by about 3 m of materials from elsewhere on the site.
Various combinations of materials for a multilayer cap are being
considered.

The Koongarra Project currently has plans for two below-grade tailings pits, each of about 14 ha. These are to be sunk 9-10 m in weathered schists, with their floors still 10-15 m above the local aquifer. The reasons for the two pits are that the first can be decommissioned after about 6 years, and rehabilitation procedures and experience obtained from that can then be used in the ultimate abandonment of the second pit.

The location of the dams is at the crest of a low interfluve divide, across the valley from the mine site, which is at the base of a Mount Brockman outlier. This is to remove the tailings ponds from any mass-movement impacts at the base of the escarpment and from the higher local precipitation found there.

The final rehabilitation has options for various combinations of capping materials which will regrade the surface to one similar to that of the original form. Best practical technology will be used to help determine the most suitable combinations of materials to be used and this will be developed, in part, by experience with the closure of the first pit.

The general question of rehabilitation has been addressed by the Environmental Section of the Australian Atomic Energy Commission in a series of options for a model using a basic turkey's nest design. Various caps for this model have been considered, together with the rates of their likely loss of integrity. For Ranger such a structure would involve a containment of 110 ha, with outer walls graded down to 1 in 6 slopes and a total area of about 200 ha. The difference in elevation between the top of the capped tailings and the top of the dam walls is sufficient to create a hollow upper surface to the man-made landform. Although ponding of some water may occur in the wet season, it will never overflow, while evaporation exceeds pre-cipitation. However, in the dry season, cap materials will dry out and any vegetation will die off, leaving the finer materials liable

to removal by deflation. It has been estimated that wind action could
remove about 0.5 mm/yr. At this rate and a uniform rate of lowering,
the cap would survive 6000 years. This rate is thought to be
greater than the denudation that would occur on the 9° side slopes.
However, little is known as yet on the likely rates of erosion of un-
vegetated or vegetated, man-made slopes, up to 30 m high and 180 m
long in this wet and dry tropical climate. It seems probable that
disturbed slopes at such gradients could erode fairly quickly,
particularly if slopes could not be stabilised by vegetation and rip-
rap in vulnerable locations. It is probable that experimentation on
trial slopes will establish likely rates of denudation and that these
will influence the final planning for surface rehabilitation. If it
can be demonstrated that the integrity of such a structure could
survive for several thousands of years, it may be used as an alternative
to replacing tailings in the mine pit.

In the summary data from the Commission, likely pathways for waterborne pollutants were also considered within the present regime, and
for future regimes involving CO_2 heating and glaciation. In the absence
of real data, calculations have been made with the best information and
estimates available. More work: using tracers to simulate hydrological
dispersion of polluted waters in the riverine, backwater swamps and
estuarine-tidal environments; using sediment sampling procedures at
gauging stations throughout the system and throughout the wet season;
and assessing sedimentation on the flood plains, as well as throughput
sediment loads, will help provide more realistic figures for such
calculations. To date, this information has been required not only to
make assessments of the fate of dams after abandonment, but also to help
the A.A.E.C. calculate local, regional and global risks to health now
and in the long-term future.

ANNEX II

SUMMARY OF GEOMORPHOLOGICAL CONSIDERATIONS
ASSOCIATED WITH THE SITING OF URANIUM MILL TAILINGS DISPOSAL DAMS
AND THEIR ABANDONMENT AFTER DECOMMISSIONING AND REHABILITATION

Dr. R.F. WARNER

The disposal of uranium mill tailings on or just below the land surface poses several major problems to the environment.

Tailing dumps or dams represent above background concentration at or near the surface of heavy metals, radioactive materials and other noxious and perhaps unstable chemical compounds, often in large volumes (at Ranger, Northern Territory, Australia, for example, the volume of tailings will be between 28 and 55 mill m³).

The dump or dam represents a disturbance to the integrity of the natural site, where previously geomorphological processes may have been slowly modifying natural forms in some kind of dynamic equilibrium. The area disturbed may be as little as a few hectares to more than a square kilometre.

The ultimate man-made landform is a new, unconsolidated surface of unknown stability, particularly in terms of responses to contemporary geomorphological processes. The landform is frequently above grade and therefore represents some form of additional surface loading.

Consequently it is necessary to find sites in the mine locality which are relatively stable, which can accommodate large volumes of tailings above or below grade, and which can be safely rehabilitated for long periods of time, without greatly accelerating natural geomorphological processes. Loss of integrity through structural failure could cause long-lasting pollution of surface and/or ground-water environments.

Thus the site has to be considered at three stages:

a)   in the actual selection of the site for storage of tailings,
b)   in the maintenance of structural integrity and safety during the operation of the mine, and
c)   in site rehabilitation, for safe and long-lasting abandonment, with low maintenance.

Tailings are discharged in various states, ranging from dry to wet in pits or dams. The size of these may range from a relatively small trench, to dams, rock basins or large turkey's nest dams. Rock basins have been favoured in the deranged drainage of the glaciated Canadian shield. A variety of tailings management schemes is proposed in Australia including turkey nest dams and below-grade disposal.

The size of the dump or dam required affects site selection, its potential stability and rehabilitation. Problems tend to increase with increasing size of the structure required.

1.   The geomorphological concerns are with individual landforms, their positions in slope sequences and with the nature of adjacent channel reaches.

Many individual landforms may be part of some relict landscapes and may therefore be dateable. The age of the landform can often be established using traditional technologies (morphology, stratigraphy,

pedology, pollen and other inclusions). Sometimes, this may be checked
using Carbon 14 or other absolute dating techniques. In these ways it
is possible to confirm the relative stability of the sites for several
thousands of years or longer.

The following examples are given to demonstrate the possibilities.

a) An erosion surface may be millions of years old (the late
   Tertiary surface - Koolpinyah - in Northern Australia).
   However it is necessary to consider the degree of truncation
   in lateritic soils, as well as the role of deep weathering.
   These may indicate more recent modifications.

b) A structural surface, such as the cap rock of a mesa, may be
   very old and being consumed slowly by retreating freeface
   slopes of the escarpment. In such an environment, rainfall
   infiltration may be high and consequently ground water movements to springs and seepage zones could cause some problems.

c) Pediments, water-cut surfaces of considerable antiquity, are
   frequently regarded as very stable forms. However, these
   gravel-veneered surfaces are often dissected by arroyos,
   indicating more recent stages of rejuvenation. These may be
   attributable to climatic and/or land use changes. Their use
   in American examples has already been advocated.

d) River terraces mark locations of surviving remnants of earlier
   flood plains. Survival is often dependent on having protective
   bedrock spurs and wide valleys. Since these represent the
   last few stages in the fluvial sequence, they are often younger
   than pediments and typical age ranges may be from 3/4000 to
   30/40,000 years. Ages have been established in many areas and
   these often relate to climatic and sea-level changes of the
   Pleistocene and Holocene. Marginal dissection, ground-water
   movements and inundation by extreme events may pose problems.
   The areas of terrace surfaces are limited and their survival
   in narrow valleys is unlikely.

e) Alluvial fans and talus slopes are steeper land forms of
   depositional origin. They may be subject to natural burial by
   deposits derived from up valley or up slope, but they can be
   dissected and the movement of ground water in the unconsolidated
   materials could be a problem.

f) Flood plains are active parts of the contemporary high-flow
   channel. As such, as was shown in the Gallup example, they
   should be avoided. Channels and their adjacent flood plains
   are not very permanent features. They are subject to periodic
   inundation and channel migration can remove flood plains, as
   well as part of adjacent terraces.

The landforms described above are among the flatter components of slope
sequences. They may be found above or below scarps (active slope zones),
below valley walls or separated by minor bluffs. The landscape is in
effect a continuum of slopes from divides to valley floors, with steeper
erosional (more unstable) elements and flatter, often depositional
(more stable) elements. Consequently, for landform stability, it is
necessary to consider flatter slope elements, as well as rates of
development up- and down-slope. To examine these, it requires some
observations, not only of age, but also of:

a) slope geology (lithology and structure - for mass movement
   phenomena) and soil characteristics,

b) slope hydrology, infiltration and runoff characteristics,

c) slope vegetation and landuse, which may influence stability,

d) slope angle and length, providing gradient and up-slope
   catchment length, and

e) slope denudation, and hazards as determined from relict
   features and from contemporary processes.

A whole range of technology is available for such investigations.

Although disposal sites should not normally be established near major rivers, large sites frequently involve smaller channels or parts of the drainage network or may be near channel systems. Thus it is important to consider elements of channel instability and other hazards.

Channel instability can be demonstrated with recent bank erosion evidence, through many meander cutoffs on the present flood plain, with braiding where there was once a meandering channel. Historical evidence can be searched through old photographs, maps and air-photographs. Stratigraphical, pedological, vegetation, pollen and Carbon 14 evidence can be studied.

Long-profile instability can be investigated through the examination of nearby headcuts or nickpoints, and of migrating slugs of sediment which may be moving down the channel. In many cases it is possible to investigate the channel's responses to climatic and land-use changes and the evidence for extreme events can be mapped and dated, in order to show the impacts of infrequent high flows.

In near-coast environments, the impacts of storm surges, those of low and high sea levels in the last few thousand years, and the effects of flooding need to be considered from surviving landforms and sub-alluvial fills and their stratigraphics.

Geomorphological hazards may include slope and channel forms. Both may be close to some threshold where a major change may occur. Other hazards include those induced by tectonic events, climatic, sea-level and man-induced changes. Where such changes have occurred in the past, they can be interpreted from the geomorphological or landform record. Future impacts may then be assessed from the local evidence or by the use of analog situations.

Much evidence and technology relevant to site selection has been presented in the Highway Bridge Report[*]

2. Once adequate safeguards have been considered in site selection, the maintenance of the tailings dump or dam during the working life of the mine should be no more than an engineering problem. However, exceptions have occurred with a dam failure at Gallup and a small overflow at Ranger following a breach which had to be made in a high rain storm to prevent damage to earthworks.

3. Finally, the rehabilitation of the tailings site becomes the ultimate problem, that of trying to guarantee the integrity of the structure for a few thousand years.

    The siting of dams or dumps above grade assumes that grade can be loaded without greatly increasing the rate of denudation which existed before the structure was built. This assumption can never be quite true, because grade represents the contemporary natural equilibrium surface between endogenic and exogenic forces. The hope is that any accelerated increase in the denudation rate is 'acceptable', in that the slopes and cap materials of structures will survive for a few thousand years at least, and therefore, radio-active materials and other matter will be contained for a long time.

    The siting of tailings below grade is logically more acceptable, if interactions between waste materials and ground water can be minimised, and if the man-made rehabilitated surface, including the caps, can be made similar to the natural graded surface. There are problems with this in terms of materials, riprap surfacing and re-establishing vegetation.

---

* Shen et al. "Methods for Assessment of Stream-Related Hazards to Highways and Bridges". (see Bibliography).

The surface stability of the man-made landforms is very important in assessing the structural integrity of the pile. If it is possible to reconstruct a surface which is similar to the natural surface, and the latter was relatively stable, there is hope. If not, experimentation at various sites may have to precede closure and abandonment. Various surface types of rehabilitation can be tested during the life of the mine to determine likely rates of subsequent denudation.

The earlier assessments of channel and slope stability, as well as hazards, will be as relevant after closure and abandonment, as they were in site selection.

GEOMORPHIC ASSESSMENT OF URANIUM MILL TAILINGS DISPOSAL SITES

SUMMARY REPORT OF THE WORKSHOP BY THE PANEL OF GEOMORPHOLOGISTS

Professor S.A. SCHUMM
Department of Earth Resources, Colorado State University, USA

Dr. J.E. COSTA
Department of Geography, University of Denver, USA

Dr. T. TOY
Department of Geography, University of Denver, USA

Dr. J. KNOX
Department of Geography, University of Wisconsin, USA

Dr. R. WARNER
Department of Geography, University of Sydney, Australia

Dr. J. SCOTT
Director, Terrain Sciences Division,
Geological Survey of Canada

ABSTRACT

The following report of the panel of geomorphologists is a summary of the principal findings of the geomorphological Workshop with respect to its three objectives :

1) examination of geomorphic controls on site stability,

2) demonstration of the application of the principles of geomosphology to the siting (and design) of stable tailings disposal containment systems,

3) development (in outline) of a procedure for the evaluation of long-term stability of tailings disposal sites.

The purpose of the workshop and the deliberations of the panel was to consider geomorphic hazards to the long-term disposal of radioactive materials, in particular uranium mill tailings. Within a landscape there are locations which are favorable for long-term disposal of such radioactive materials and there are others that are potentially very hazardous. Hazardous locations may be geomorphicly unsuitable because of relatively rapid landform changes or because the location is subject to inundation by extreme hydrologic events.

The very nature of the landscape and its components (hill-slopes, drainage networks, rivers) provide an indication of site stability as do Pleistocene and Holocene alluvial deposits that reflect the effect of climate fluctuations and extreme hydrologic events.

A uranium-tailings disposal site will be located within the landform complex that is termed the fluvial system (Fig. 1). The fluvial system can be divided into three zones as follows :

Zone 1 - the drainage basin, a sediment and runoff production zone

Zone 2 - the major streams, a sediment and water transport zone

Zone 3 - the piedmont or coastal plain, a depositional zone.

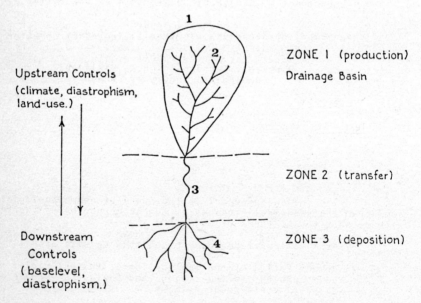

Figure 1. Sketch of fluvial system showing location of tailings disposal sites.

1) divide or plateau
2) valley head or tributary
3) major valley, flood plain, terrace
4) alluvial plain, fan, delta, pediment

Obviously this tripartite division of a complex system is far too simplistic, and in nature a clear distinction between them is difficult. In addition, there can be sediment deposition and storage in zones 1 and 2 and erosion in zone 3 as these zones adjust to climate change, tectonic effects, and base-level change. Nevertheless, the subdivision does aid in the discussion of a tailings-site location, and the geomorphic hazards that influence site stability. For example, the sites will be located differently in each zone as follows (Fig. 1):

> Zone 1 - drainage divide or plateau sites, valley head or tributary sites

> Zone 2 - major valley sites, terraces and flood plains

> Zone 3 - plain sites, bajada, alluvial fan, coastal plain, delta and pediment.

If a landscape is static then only the effect of extreme hydrologic events need to be considered in site selection. Clearly, if landform evolution is slow or if the time span of concern is short, then most variables that cause landform change can be dismissed. Unfortunately this is not the case and in Table 1 the variables that influence the fluvial system are listed in order of their significance. Time, of course, is not a true independent variable, but it is measure of the extent of landform evolution under the influence of geomorphic agents of erosion and deposition. The initial relief, geology and climate establish the rate at which change will occur and the process that will act (fluvial, eolian or glacial). Climate, through its control of vegetation, relief through its control of the gravitational forces acting, and geology through its control on erodibility and water loss, will determine runoff and sediment yield per unit area of zone 1. The erosional forces generated by flowing water acting of the geologic materials will develop a characteristic drainage network and hillslope morphology, which in turn will influence the quantity of water and sediment delivered to zones 2 and 3. The quantity and type of sediment and the hydrologic character of the runoff events will, in the absence of other external controls (baselevel and tectonic effects), establish the channel, valley, and depositional landform morphology of zones 2 and 3.

Table 1. Fluvial System Variables

1	Time
2	Initial Relief (tectonics)
3	Geology (lithology and structure)
4	Climate
5	Hydrology (runoff and sediment yield)
6	Drainage network morphology
7	Hillslope morphology
8	Hydrology (discharge of water and sediment to zones 2 and 3)
9	Channel and valley morphology (zone 2)
10	Depositional system morphology (zone 3)

It should be clear why the landform complex is referred to as a system because of the interaction of independent and dependent variables throughout the fluvial system. Upstream hydrologic changes will influence the landform morphology and stability of zones 2 and 3, and a baselevel change in zone 3 can have an effect upon zones 1 and 2.

Of the variables listed in Table 1 the following appear to be most significant for landform stability :

1) time - natural landform evolution rate and processes

2) relief - change of relief as a result of tectonics (uplift or subsidence) and baselevel change (rise or fall of sea level, lake level, or abrupt change of stream profile)

3) climate change - change of vegetation cover and type and hydrologic regimen.

In addition to these three categories of variables, changes of land use, and deforestation by man, fire or disease will also have a significant influence on the fluvial system.

The effect of these changes can be expressed by changes of runoff (discharge), sediment yield, and baselevel. Depending on the direction of change, the result can be beneficial or detrimental to tailings disposal sites in all three zones of the fluvial system.

In Tables 2, 3 and 4, the effect of climate change, tectonics and land use changes upon the variables runoff, sediment yield, and baselevel are qualitatively indicated by plus, minus, or zero notation, thereby indicating respectively an increase, decrease, or no change.

Table 5 is a list of geomorphic hazards for each zone of the fluvial system ; in each zone changes can occur in the erosional and depositional regime and in the drainage network or river pattern. Depending on the magnitude of the change of the independent variables the geomorphic hazard can be extreme or moderate. For example, deposition or erosion in a stream channel (see Table 5, zone 2, items 3a1 and 3b1) can be minor or great depending on the change of sediment supply and runoff. Some hazards are essentially local in effect such as meander growth and cutoffs (see Table 5, zone 2, items 3c1, 3c3).

The next step in the analysis is to relate the variables to the hazards in order to provide at least a qualitative assessment of site stability. This analysis will be undertaken and the results will be presented at the IAEA/NEA International Symposium on the Management of Wastes from Uranium Mining and Milling which is to be held in Albuquerque, NM, USA, May 10-14, 1982.

As noted above, even if there is no geomorphic hazard the location of a site in a valley renders it susceptible to extreme flood events. Normally hydrologic records are too short to provide information on extreme events. Therefore, a geomorphic approach to this problem is necessary because in dealing with long time spans, the possibility of rare but extreme geomorphic events such as catastrophic floods cannot be ignored. For example, Table 6 shows that the chance of a 1000-year flood occurring in any single year is slight, only 0.1 percent. But in a 1000-year period the chance of the 100-year flood occurring is 63 percent, and in 5000 years the chance is near certainty, over 99 percent. An even more severe flood (5000-year flood) has an 18 percent chance of occurring in 1000 years, which is not an insignificant risk. Conventional engineering analysis has a great deal of trouble identifying the magnitude of a 1000-year flood as well as the recurrence intervals of large catastrophic floods. Geomorphic techniques offer an alternative solution to these questions.

Table 2. Climate Change

Variables		arid to semiarid (a) semiarid to arid (b)	semiarid to subhumid subhumid to semiarid	subhumid to humid humid to subhumid	nonglacial to glacial	nonglacial to periglacial	uniform to seasonal
Runoff	a	+	+	+	+	+	-
	b	-	-	-			
Sediment Yield	a	+	-	r	+	+	+
	b	-	+	+	+	+	
Baselevel							
lake	a	+	+	+	+	+	0
	b	-	-	-			
ocean		0	0	0	-	0	0

- 73 -

Table 3. Tectonics

Variables	Upstream		On Site		Downstream	
	Up	Down	Up	Down	Up	Down
Runoff	0	0	0	0	0	0
Sediment Yield	+	-	+	+	0	0
Baselevel	0	0	+	-	+	-

Table 4. Land Use and Vegetation Change

Variables	fire	increased agriculture	timbering	Dam Construction Upstream	Downstream	Gravel Mining Upstream	Downstream	Channelization Upstream	Downstream
Runoff	+	+	+	-	0	0	0	+	0
Sediment Yield	+	+	+	-	0	- or +	0	+	0
Baselevel	0	0	0	0	+	0	0	0	-

Table 5. Geomorphic Hazards

<u>Zone 1   Landforms</u>   (Fig. 1)

1. <u>Drainage Networks</u>
    a) Erosion
        1. rejuvenation - incision of all channels and headward
            lengthening into undissected areas of the drainage
            basin.
        2. extension - headward growth of the drainage network
            without rejuvenation.

    b) Deposition
        1. valley filling - major deposition causing significant
            elevation of valley floor.

    c) Pattern change
        1. capture - diversion of runoff and sediment from one
            channel to another with major channel adjustments
            (see 3a,b,d below).

2. <u>Slopes</u>
    a) Erosion

        1. denudation - normal slope erosion and retreat
        2. dissection - incision of slope by rills, gullies, or
            by drainage network extension (1a2 above)
        3. mass failure - all forms of mass movement, slump,
            landslide, etc.

<u>Zone 2   Landforms</u>   (Fig. 1)

- 3. <u>Channels</u>
    a) Erosion
        1. degradation - general lowering of stream bed by incision
        2. nickpoint formation and migration - development of
            break in longitudinal profile that migrates upstream
        3. bank erosion - channel widening or local erosion as
            related to meander shift.

    b) Deposition
        1. aggradation - general raising of stream bed by deposition
        2. back and downfilling - deposition in channel that migrates
            upstream or downstream
        3. berming - narrowing of channel by deposition of fine
            sediments on banks

Table 5. Geomorphic Hazards (Cont'd)

    c) Pattern change

        1. meander growth and shift - increase of meander amplitude and downstream migration
        2. island and bar formation and shift - bars form by local deposition and become islands when colonized by vegetation; islands as a result of in-channel erosion and deposition can shift position
        3. cutoffs - neck and chute cutoffs of meanders locally steepen gradient
        4. avulsion - major shift of channel position

    d) Metamorphosis - a complete change of channel morphology as expressed by pattern change

        1. straight to meandering - development of bends and probable increase of width; gradient decreases
        2. straight to braided - development of bars and islands, increase of width and reduction of depth, may reflect aggradation
        3. braided to meandering - narrowing, deepening, and development of bends, decrease of gradient
        4. braided to straight - narrowing and deepening
        5. meandering to straight - increase of gradient
        6. meandering to braided - widening, shallowing, increase of gradient, may indicate aggradation.

Zone 3 Landforms (Fig. 1)

  4. Piedmont and Alluvial Plain (alluvial fan, delta, bajada, pediment)

    a) Erosion

        1. dissection - development of channels and incision of surface

    b) Deposition

        1. aggradation - general filling of channels and deposition on surface of fan or delta
        2. progradation - lengthening of delta or fan with upstream deposition

    c) Pattern change

        1. rejuvenation and extension - incision of existing channels and headward growth
        2. development of drainage pattern - growth of drainage network on undissected surface
        3. avulsion - channel shift and bifurcation on fan or delta
        4. capture - diversion or piracy of water and sediment leading to rejuvenation of captor stream.

Table 6

PROBABILITY OF EXTREME EVENTS

$$q = \frac{1}{R.I.} \qquad\qquad q = 1 - \left[1 - \frac{1}{R.I.}\right]^n$$

n = next n years
R.I. = vecurrence interval
q = probability of occurrence

R.I.	In any given yr	In 10 yr	100 yr	1000 yr	5000 yr	10,000 yr
500 yr event	0.2 %	2 %	18 %	86.5 %	99.995 %	99.999 %
1000 yr event	0.1	1	9.5	63	99.33	99.99
5000 yr event	0.02	0.2	2	18	63	86.5

At present, the design of tailings site security from flooding is based on the concept of the probable maximum flood (PMF), which is derived from the probable maximum precipitation. The PMF was developed for situations where substantial risk of loss of life exists in the event of failure of critical structures such as dam spillways. The PMF is supposed to represent the worst possible flood conditions. No probability can be realistically attached to the PMF, and the implication is that complete protection is provided. The implication of no risk is false ; in fact the risk is unknown. On numerous occasions PMF's have been exceeded and recurrence intervals for PMF's vary over at least four orders of magnitude. PMF's are calculated from data based on the historical period of rainfall and runoff records, and thus are iterative and change as the history of large rainstorms accrues. Examples include upward adjustments of PMF values for the Cherry Creek Dam near Denver, Colorado, after a severe 1965 storm, and the Susquehanna River at Harrisburg, Pennsylvania, following Tropical Storm Agnes flooding in 1972.

The geomorphic record of landforms and deposits in a valley can give reasonable estimates of the magnitude and frequency of extreme events. Results of hydrologic reconstructions using sediments deposited by known large catastrophic floods are within 20-30 percent of accepted peak discharge values. Climate and land use do not seem to be controlling factors in determining catastrophic flood peaks in small basins in the United States ; drainage area is the most important variable. Maximum floods for basins up to 500 square miles in the United States follow the power function :

$$Q_{max} = 11,000 \ (\text{Drainage Area})^{0.61} \ (\text{cfs and sq. mi. units}).$$

Not all basins where tailings disposal sites are planned will have appropriate sedimentological and/or stratigraphical evidence of extreme floods. This could be because evidence has not been preserved, or because no such floods have occurred. However investigations in other basins in the region may well provide paleodischarge information which can be extrapolated to the site under consideration.

Recommended procedures for investigation of the history of extreme floods at a potential disposal site would be :

a) Compile all available hydrological data on past floods in the region. Compare the peaks with some limiting maximum flood curve, such as the one whose equation is given above.

b) Compute a PMF for the site and compare the estimate to the value found from the maximum flood curve. If it is less than this value, the long-term (1000-year plus) flood risk may be underestimated by the PMF discharge.

c) Investigate channel, floodplain, and low terrace exposures to look for stratigraphic evidence of prehistoric large floods. This investigation should include collection of particle size data and appropriate datable carbonaceous materials from important horizons for the purpose of recurrence estimates. These on-site investigations should be extended to surrounding watersheds to obtain a view of the regional flood history.

d) If appropriate sedimentological evidence is found, such as slack-water sediments and/or flood boulder bars, some simple surveying will be required to establish relative elevations and cross-sections. It may take as long as several months for radiocarbon samples to be processed by laboratories, but the paleohydraulic reconstructions and dates will provide an objective estimate of the long-term flood frequency and magnitude of potential sites.

e) As a general rule, floodplains and low terraces are not appropriate sites for disposal of uranium tailings. Data obtained by geomorphological techniques has indicated that historic catastrophic floods, having recurrence intervals over 1000 years, have inundated and eroded such surfaces.

The purpose of geomorphic assessment of a site is to locate the tailings disposal facility on a geomorphic surface that will be relatively stable for long periods. Each site will have its own characteristic climate, geology, and erosional stability. For the geomorphologist to identify potential geomorphic hazards the following types of information are sought for such an assessment :

1) Aerial photographs of the site and the surrounding area

2) Topographic maps of the site and the surrounding area

3) Information on

   a. Climatology

   b. Hydrology

   c. Vegetation type and density

   d. Geologic maps, reports

   e. Soils maps, reports.

4) Geomorphology

   a. Stream morphology

      - Information on stream dimensions, gradient, pattern, sediment character, longitudinal profile. Location of bedrock controls

   b. Hillslope and scarp morphology

      - Information on slope inclination, profile shape, sediment character, evidence of mass movement (creep, slump, rock falls, etc.) presence of bedrock or gravel cap rock.

   c. Drainage network morphology

      - Information (maps) showing general pattern of streams and tributaries as related to the site. A large-scale map showing the patterns over a 10 km radius from the site will be of considerable value. Observations on water bodies downstream of site. General observations of pattern instability.

Obviously much of this information may not be readily available and certainly the geomorphic information will have to be obtained from field investigations. Although this will add to the cost of site investigation, selection and development, the cost will be negligible when compared with the costs of construction, mining and tailings pile development.

BIBLIOGRAPHY OF SELECTED REFERENCES

I. SELECTED BIBLIOGRAPHY ON THE APPLICATION OF GEOMORPHOLOGICAL
    AND BOTANICAL TECHNIQUES IN FLOOD HYDROLOGY INVESTIGATIONS
    (compiled by Dr. John E. Costa)

Allen, J.R.L., 1968, Current ripples: North-Holland, Amsterdam, 433 p.

Allen, J.R.L., 1973, Phase differences between bed configuration and
    flow in natural environments, and their geologic relevance:
    Sedimentology, v. 20, p. 323-329.

Anderson, M.G., and Calver, A., 1977, On the persistence of landscape
    features formed by a large flood: Institute of British Geographers
    Transactions, v. 2, no. 2, p. 243-254.

Anderson, M.G. and Calver, A., 1980, Channel plan changes following
    large floods: in Timescales in Geomorphology, Cullingford, R.A.,
    Davidson, D.A., and Lewin, J., eds., John Wiley, N.Y., p. 43-52.

Andrews, E.D., 1980, Effective and bankfull discharges of streams in the
    Yampa River Basin, Colorado and Wyoming: Journal of Hydrology,
    v. 46, p. 311-330.

Baker, V.R., 1973, Paleohydrology and sedimentology of Lake Missoula
    flooding in eastern Washington: Geological Society of America,
    Special Paper 144, 79 p.

_____1975, Flood hazards along the Balcones Escarpment in central Texas:
    Alternative approaches to their recognition, mapping, and management:
    Austin, Texas University, Bureau of Economic Geology Circular 75-5,
    22 p.

_____1977, Stream-channel response to floods, with examples from Central
    Texas: Geological Society American Bulletin, v. 88, p. 1057-1071.

Baker, V.R., and Ritter, D.F., 1975, Competence of rivers to transport
    coarse bedload material: Geological Society of America Bulletin,
    v. 86, p. 975-978.

Balog, J.D., 1978, Flooding in Big Thompson River, Colorado, tributaries:
    Controls on channel erosion and estimates of recurrence interval:
    Geology, v. 6, p. 200-204.

Beverage, J.P., and Culbertson, J.K., 1964, Hyperconcentrations of
    suspended sediment. Journal of Hydraulics Division, American
    Society of Civil Engineers, v. HY6, p. 117-128.

Birkeland, P.W., and Shroba, R.R., 1974, The status of the concept of
    Quaternary soil-forming intervals in the western United States: in
    Quaternary Environments, edited by W.C. Mahaney, York University-
    Atkinson College Geographical Monographs, p. 241-276.

Bradley, W.C., and Mears, A.I., 1980, Calculations of flows needed to
    transport coarse fraction of Boulder Creek alluvium at Boulder,
    Colorado: Geological Society of America Bulletin, v. 91, p. 1057-
    1090.

Brink, V.C., 1954, Survival of plants under flood in the lower Fraser
    River Valley, British Columbia: Ecology, v. 35, p. 94-95.

Brunsden, D., and Thornes, J.B., 1980, Landscape sensitivity and change: Earth Surface Processes, v. 5, p. 463-484.

Burkham, D.E., 1981, Uncertainties resulting from changes in river form: Hydraulics Division, American Society of Civil Engineers, v. 107, HY5, p. 593-610.

Campbell, C.A., Paul, E.A., Rennie, D.A., and McCallum, K.J., 1967, Factors affecting the accuracy of the carbon-dating method in soil humus studies: Soil Science, v. 104, p. 81-85.

Cooley, H.E., Aldridge, B.N., and Euler, R.C., 1977, Effects of the catastrophic flood of December 1966, North Rim area, Eastern Grand Canyon, Arizona: U.S. Geological Survey Professional Paper 980, 43 p.

Costa, J.E., 1974, Response and recovery of a Piedmont watershed from Tropical Storm Agnes, June 1972: Water Resources Research, v. 10, p. 106-112.

_____1974, Stratigraphic, morphologic, and pedologic evidence of large floods in humid environments: Geology, v. 2, p. 301-303.

_____1978, Colorado Big Thompson flood: Geologic evidence of a rare hydrologic event: Geology, v. 6, p. 617-620.

_____1978, Holocene stratigraphy in flood-frequency analysis: Water Resources Research, v. 14, p. 626-632.

Council of the American Meteorological Society, 1978, Flash Floods-A National Problem: American Meteorological Society Bulletin, v. 59, p. 585-586.

Crippen, J.R., and Bue, C.D., 1977, Maximum floodflows in the conterminous United States: U.S. Geological Survey Water-Supply Paper 1887, 52 p.

Day, T.J., 1972, The channel geometry of mountain streams: in Slaymaker, O. and McPherson, J., eds., Mountain Geomorphology, p. 141-149.

Emmett, W.W., 1974, Channel changes: Geology, v. 2, p. 271-272.

Everett, B.L., 1968, Use of cottonwoods in an investigation of the recent history of a floodplain: American Journal of Science, v. 266, p. 417-439.

Follansbee, R., and Sawyer, L.R., 1948, Floods in Colorado: U.S. Geological Survey Water-Supply Paper 997, 151 p.

Fritts, H.C., 1976, Tree Rings and Climate: London Academic Press, 567 p.

Gardner, J.S., 1977, Some geomorphic effects of a catastrophic flood on the Gand River, Ontario: Canadian Journal of Earth Sciences, v. 14, p. 2294-2300.

Geyh, M.A., Benzler, J.H., Roeschmann, G., 1971, Problems of dating Pleistocene and Holocene soils by radiometric methods: in Paleo-pedology, Origin, Nature, and Dating of Paleosols, edited by D.H. Yaalon, International Society of Soil, Science, and Israel University Press, p. 63-75.

Glancy, P.A., and Harmsen, L., 1975, A Hydrologic Assessment of the September 14, 1974, Flood in Eldorado Canyon, Nevada: U.S. Geological Survey Professional Paper 930, 28 p.

Gregory, K.J., ed., 1977, River channel changes: John Wiley and Sons Inc., N.Y., 448 p.

Gupta, A., 1975, Stream characteristics in eastern Jamaica, an environment of seasonal flow and large floods: American Journal of Science, v. 275, p. 825-847.

Gupta, A., and Fox, H., 1974, Effects of high-magnitude floods on channel form: a case study from the Maryland Piedmont: Water Resources Research, v. 10, p. 499-509.

Hack, J.T. and Goodlett, J.C., 1960, Geomorphology and forest ecology of a mountain region in the Central Appalachians: U.S. Geological Survey Professional Paper 347, 66 p.

Hand, B.M., 1969, Antidunes as trochoidal waves: Journal of Sedimentary Petrology, v. 39, p. 1302-1309.

Hand, B.M., Wessel, J.M., and Hayes, M.O., 1969, Antidunes in the Mount Toby Conglomerate (Triassic), Massachusetts: Journal of Sedimentary Petrology, v. 39, p. 1310-1316.

Hardison, C.H., 1973, Probability distribution of extreme floods: in Highways and the Catastrophic Floods of 1972, Highway Research Board Report no. 479, p. 42-45.

Harms, J.C., 1969, Hydraulic significance of some sand ripples: Geological Society of America Bulletin, v. 80, p. 363-396.

Harrison, S.S., and Reid, J.R., 1967, A flood-frequency graph based on tree-scar data: Proceedings, North Dakota Academy of Science, v. 21, p. 23-33.

Helley, E.J., and LaMarche, V.C., 1973, Historic flood information for northern California streams from geological and botanical evidence: U.S. Geological Survey Professional Paper 485-E, 16 p.

Hickey, J.J., 1969, Variations in low-water streambed elevations at selected stream-gaging stations in Northwestern California: U.S. Geological Survey-Water Supply Paper 1879-E,

Holmes, R.L., Stockton, C.W., and LaMarche, V.C., 1979, Extension of river flow records in Argentina from long tree-ring chronologies: Water Resources Bulletin, v. 15, p. 1081-1085.

Jahns, R.H., 1947, Geologic features of the Connecticut Valley, Massachusetts as related to recent floods: U.S. Geological Survey Water-Supply Paper 996, 158 p.

Karcz, I., 1972, Sedimentary structures formed by flash floods in southern Israel: Sedimentary Geology, v. 7, p. 161-182.

Karcz, I., 1969, Mud pebbles in a flash flood environment: Journal of Sedimentary Petrology, v. 39, p. 333-337.

Kelsey, H.M., 1980, A sediment budget and an analysis of geomorphic process in the Van Duzen River Basin, north coastal California, 1941-1975: Geological Society of America Bulletin, Part II, v. 91, p. 1119-1216.

Kennedy, J.F., 1961, Stationary waves and antidunes in alluvial channels: California Institute of Technology, W.M. Keck Lab. Hydraul. Water Resour., Report KH-R-2:146 p., Pasadena, Calif.

Knott, J.M., 1971, Sedimentation in the Middle Fork Eel River Basin, California: U.S. Geological Survey Open File Report, Menlo Park, Ca.

Knox, J.C., 1979, Geomorphic evidence of frequent and extreme floods: Improved Hydrologic Forecasting - why and how: American Society of Civil Engineers, p. 220-238.

Kopaliani, Z.D., and Romashin, V.V., 1970, Channel dynamics of mountain rivers: Soviet Hydrology, v. 5, p. 441-452.

Laing, D., and Stockton, C.W., 1976, Riparian dendrochronology: A method for determining flood histories of ungaged watersheds: Report to Office of Water Research and Technology, OWRT project #A-058-ARl7, 18 p. (Lab of Tree-Ring Research, University of Arizona, Tucson, Arizona).

Lock, W.W., Andrews, J.T., and Webber, P.J., 1979, A manual for lichenometry: British Geomorphological Research Group, Technical Bulletin no. 26, 47 p.

Malde, H.A., 1968, The Catastrophic Late Pleistocene Bonneville Flood on the Snake River Plain, Idaho: U.S. Geological Survey Professional Paper 596, 52 p.

Matthes, G.H., 1947, Macroturbulance in natural streamflow: American Geophysical Union Transactions, v. 28; p. 255-262.

McGowen, J.H., and Garner, L.E., 1970, Physiographic features and stratification types of coarse-grained point bars: modern and ancient examples: Sedimentology, v. 14, p. 77-111.

McKee, E.C., Crosby, E.J., and Berryhill, H.C., 1967, Flood deposits, Bijou Creek, Colorado, June 1965: Journal of Sedimentary Petrology, v. 37, p. 829-851.

Middleton, G.V., ed., 1965, Primary sedimentary structures and their hydrodynamic interpretation: Society of Economic Paleontologists and Mineralogist Special Publication no. 12, 265 p.

Middleton, G.V., 1965, Antidune cross-bedding in a large flume: Journal of Sedimentary Petrology, v. 35, p. 922-927.

Newson, M., 1980, The geomorphological effectiveness of floods - a contribution stimulated by two recent events in mid-Wales: Earth Surface Processes, v. 5, p. 1-16.

Patton, P.C., 1977, Geomorphic criteria for estimating the magnitude and frequency of flooding in central Texas: unpublished Ph.D. dissertation, Austin, Texas University, 225 p.

Patton, P.C. and Baker, V.R., 1976, Morphometry and floods in small drainage basins subject to diverse hydrogeomorphic controls: Water Resources Research, v. 12, p. 941-952.

Patton, P.C., and Baker, V.R., 1977, Geomorphic response of central Texas stream channels to catastrophic rainfall and run-off: in Geomorphology of Arid and Semi-arid Regions, edited by D. Doehring, State University of New York Publications in Geomorphology, p. 189-217.

Phipps, R.L., 1970, The potential use of tree rings in hydrologic investigations in eastern North America, with some botanical considerations: Water Resources Research, v. 6, p. 1634-1640.

Reckendorf, F.F., 1973, Techniques for identification of flood plains in Oregon: unpublished Ph.D. dissertation, Oregon State University, Corvallis, Oregon, 344 p.

Riggs, H.C., 1974, Flash flood potential from channel measurements: International Association of Scientific Hydrology, Symposium of Paris, IASH-AISH Publication no. 112, p. 52-56.

Ritter, D.F., 1975, Stratigraphic implications of coarse-grained gravel deposited as overbank sediment, southern Illinois: Journal of Geology, v. 83, p. 645-650.

Ritter, J.R., 1974, The effects of the Hurricane Agnes flood on channel geometry and sediment discharge of selected streams in the Susquehanna River Basin, Pennsylvania: U.S. Geological Survey Journal of Research, v. 2, p. 753-761.

Schick, A., 1974, Formation and obliteration of desert stream terraces - a conceptual analysis: Zeitschrift fur Geomorphologie, v. 21, p. 88-105.

Scott, K.M., 1973, Scour and fill in Tijunga wash - a fan head valley in urban southern California - 1969: U.S. Geological Survey Professional Paper 732-B.

Scott, K.M., 1971, Origin and Sedimentology of 1969 debris flows near Glendora, California: U.S. Geological Survey Professional Paper 750-C, p. C242-C247.

Scott, K.M., and Gravlee, G.C., 1978, Flood Surge on the Rubicon River, California: Hydrology, Hydraulics, and Boulder Transport: U.S. Geological Survey Professional Paper 422-M, 40 p.

Selby, M.J., 1974, Dominant geomorphic events in landform evolution: International Association of Engineering Geologist Bulletin, v. 9, p. 85-89.

Sigafoos, R.S., 1964, Botanic evidence of floods and flood plain deposition: U.S. Geological Survey Professional Paper 485-A, 35 p.

Simons, D.B., Al-Shaikh-Ali, K.S., and Li, R., 1979, Flow resistance in cobble and boulder riverbeds: Hydraulics Division, Amer. Society Civil Engr. v.   , HY5, p. 477-488.

Soule, J.M., Rogers, W.P., and Shelton, D.C., 1976, Geologic hazards, geomorphic features, and land-use implications in the area of the 1976 Big Thompson flood, Larimer County, Colorado: Environmental Geology 10, Colorado Geological Survey, Denver, Colorado.

Stephens, M.A., Simons, D.B., and Richardson, E.V., 1975, Non-equilibrium river form: Journal of Hydraulics Division, American Society of Civil Engineers, v. HY5, p. 557-566.

Stewart, J.H., and LaMarche, V.C., 1967, Erosion and deposition in the flood of December 1964 on Coffee Creek, Trinity County, California: U.S. Geological Survey Professional Paper 422-K, 22 p.

Stockton, C.W., 1975, Long-term streamflow records reconstructed from tree rings: Papers of the Laboratory of Tree-Ring Research, No. 5, Tucson University of Arizona Press, 111 p.

Stockton, C.W., 1977, Interpretation of past climatic variability from paleoenvironmental indicators: in Climate, Climate Changes, and Water Supply, National Academy of Sciences, Studies in Geophysics, p. 34-45.

Stuiver, M., and Suess, H.E., 1966, On the relationship between radio-carbon dates and true sample ages: Radiocarbon, v. 8, p. 534-540.

Williams, G.E., 1970, The central Australian stream floods of February - March 1967: Journal of Hydrology, v. 11, p. 185-200.

Williams, G.E., 1971, Flood deposits of the sand-bed ephemeral streams of central Australia: Sedimentology, v. 17, p. 1-40.

Williams, G.P., and Guy, H.P., 1973, Erosional and depositional aspects of Hurricane Camille in Virginia, 1969: U.S. Geological Survey Professional Paper 804, 80 p.

Wolman, M.G., and Eiler, J.P., 1958, Reconnaissance study of erosion and deposition produced by the flood of August 1955 in Connecticut: American Geophysical Union Transactions, v. 39, p. 1-14.

Wolman, M.G., and Gerson, R., 1978, Relative scales of time and effectiveness of climatic in watershed geomorphology: Earth Surface Processes, v. 3, p. 189-208.

Woolley, R.R., 1946, Cloudburst floods in Utah, 1850-1938: U.S. Geological Survey Water-Supply Paper 994, 128 p.

Yalin, M.S., 1964, Geometrical properties of sand waves: Journal Hydraulics Division, American Society of Civil Engineers, v. 90, no. HY5, p. 105-119.

II. SELECTED REFERENCES ON HILLSLOPE AND SCARP MORPHOLOGY

Ahnert, F., 1960. The influence of Pleistocene climates upon the morphology of cuesta scarps on the Colorado Plateau : Annals, Assoc. Am. Geog., Vol. 50, No. 2, p. 139-156.

Carson, M.A., and Kirkby, M.J., 1972. Hillslope form and process : Cambridge, University Press. 475 p.

Gilley, J.E., Gee, G.W., Bauer, A., Willis, W.O., and Young, R.A., 1976. Runoff and erosion characteristics of surface-mined sites, in western North Dakota : Am. Soc. Agric. Eng. Paper No. 76-2027 p. 697-700 ; 704.

Kirkby, M.J. (ed.), 1978. Hillslope Hydrology : N.Y., John Wiley and Sons, 389 p.

Lusby, G.C. and Toy, T.J., 1976. An evaluation of surface-mine spoils area restoration in Wyoming using rainfall simulation : Earth Surface Processes, Vol. 1, No. 4, p. 375-386.

Ringen, B.H., Shawn, L.M., Hadley, R.F., Hinkley, T.K., 1979. Effects on sediment yield and water quality of a nonrehabilitated surface mine in North Central Wyoming : U.S. Geological Survey, Water Resources Investigations 79-47, 23 p.

Schumm, S.A., 1956. The role of creep and rainwash on the retreat of badland slopes, Am. Jour. Sci., Vol. 254, p. 693-706.

Schumm, S.A., and Chorley, R.J., 1966. Talus weathering and scarp recession in the Colorado Plateaus : Zeit. Geomorph., Vol. 10, p. 11-36.

Toy, T.J., (ed.), 1977. Erosion : research techniques, erodibility and sediment delivery : Norwich, England, Geo Abstracts, Ltd., 86 p.

Toy, T.J., 1981. Precipitation variability and surface-mine reclamation in the Green, Powder and San Juan River Basins : Jour. Appl. Meteorology, Vol. 20, No. 7,

Young, A., 1972. Slopes : London, Longman Group Ltd., 288 p.

III. MISCELLANEOUS REFERENCES

Shen, H.W., Schummn S.A., Nelson J.D., Doehring D.O., Skinner, M.M., Smith, G.L. "Methods for assessment of stream-related hazards to highway bridges". Report No. FHWA/RD-80/160, March 1981, prepared for U.S. Federal Highway Administration.

Patton, P.C., Baker, V.R., and Kochel, R.C., 1979. Slack-water deposits; a geomorphic technique for the interpretation of fluvial paleo-hydrology : in Rhodes, D.D. and Williams, G.P. eds., Adjustments of the Fluvial System : Kendall-Hunt Publishing Co., Dubuque, Iowa, p. 225-253.

Schultz, G.A., 1979, General Report Topic 1 : New methods of hydro-logical computations used for the design of water resources projects : Proc. International Symp. on Specific Aspects of Hydrological Computations for Water Projects, Leningrad, USSR.

Sokolov, A.A., 1979. The state and prospects for development of the methods of hydrological computation for water projects : Proc. International Symp. on Specific Aspects of Hydrological Computations for Water Projects, Leningrad, USSR.

AUSTRALIA - AUSTRALIE

BURGESS, P., McMahon, Burgess & Yeates, Chatswood Plaza,
    Chatswood, NSW 2067

DAVY, D.R., Environmental Sciences Division, Australian Atomic
    Energy Commission, Research Establishment, Lucas Heights, P.M.B.,
    Sutherland, NSW 2232

FRY, R.M., Supervising Scientist, Office of the Supervising
    Scientist, P.O. Box 387, Bondi Junction, NSW 2022

WARNER, R.F., Department of Geography, University of Sydney,
    Sydney, NSW 2006

CANADA

BRAGG, K., Senior Division Officer, Atomic Energy Control Board,
    270 Albert Street, Ottawa, Ontario K1P 5S9

CAMPBELL, M.C., Manager, Extraction Metallurgy Laboratory, CANMET,
    555 Booth Street, Ottawa, Ontario K1A OG1

CHAKRAVATTI, J.L., Denison Mines Limited, P.O. Box B-2600, Elliot
    Lake, Ontario P5A 2K2

LUSH, D., Manager, Aquatic Ecologist, Beak Consultants Limited,
    6870 Goreway Drive, Mississauga, Ontario L4V 1L9

OSBORNE, R.V., Head, Environmental Research Branch, Atomic Energy
    of Canada Limited, Chalk River Nuclear Laboratories, Chalk
    River, Ontario KOJ 1JO

SCOTT, J., Director, Terrain Sciences Division, Geological Survey
    of Canada, 601 Booth Street, Ottawa, Ontario K1A OE8

VASUDEV, P., Environmental Protection Service, Environment Canada,
    Place Vincent Massey, 16th Floor, Ottawa, Ontario K1A 1C8

FEDERAL REPUBLIC OF GERMANY - RÉPUBLIQUE FÉDÉRALE D'ALLEMAGNE

SCHNEIDER, B., Urangesellschaft mbH, Bleichstrasse 60-62,
    D-6000 Frankfurt am Main 1

SWEDEN - SUÈDE

OSIHN, Å., LKAB, Box 26044, S-100 41 Stockholm

UNITED STATES - ÉTATS-UNIS

COSTA, J.E., Department of Geography, University of Denver, Denver, Colorado 80208

GROELSEMA, D.H., Manager, Uranium Mill Tailings, Remedial Action Program, US Department of Energy, MS B-109, Washington DC 20545

HAMILL, K., Uranium Recovery Licensing Branch, US Nuclear Regulatory Commission, Washington DC 20555

KNOX, J., Department of Geography, University of Wisconsin, Madison, Wisconsin 53706

LICHTMAN, Criteria and Standards Division, Office of Radiation Programs, US Environmental Protection Agency, ANR-460, Washington DC 20460

McKIERNAN, J., Division Supervisor, Project Engineering Division 4542, Sandia National Laboratories, Albuquerque, NM 87185

MATTHEWS, M.L., Lead Project Engineer, Uranium Mill Tailings Remedial Actions Program, US Department of Energy, Albuquerque Operations Office, P.O. Box 5400, Albuquerque, NM 87115

NELSON, J., Civil Engineering Department, Colorado State University, Fort Collins, Colorado 80523

OVERMYER, Manager, Nuclear and Advanced Energy Systems, Ford Bacon & Davis (Utah), P.O. Box 8009, Salt Lake City, Utah 84108

PATTON, W.M., Texas Department of Health, Austin, Texas

ROGERS, V.C., Chief Scientist, Rogers & Associates Engineering Corporation, 515 East 4500 South, Suite G-100, Box 330, Salt Lake City, Utah 84110

RYON, A.D., Nuclear Fuel Cycle Program, Oak Ridge National Laboratory, P.O. Box X, Oak Ridge, Tennessee 37830

SCARANO, R.A., Leader, Uranium Mill Licensing Section, US Nuclear Regulatory Commission, MS 396-SS, Washington DC 20555

SCHUMM, S., Earth Resources Department, Colorado State University, Fort Collins, Colorado 80523

TOY, T., Department of Geography, University of Denver, Denver, Colorado 80208

VOLPE, R.L., R.L. Volpe & Associates, 110 Atwood Court, Los Gatos, California 95030

OECD NUCLEAR ENERGY AGENCY
AGENCE DE L'OCDE POUR L'ÉNERGIE NUCLÉAIRE

RAFFERTY, P.J., Radiation Protection and Waste Management Division, OECD Nuclear Energy Agency, 38 Boulevard Suchet, 75016 Paris, France

WORKSHOP ON

URANIUM ORE PROCESSING/TAILINGS CONDITIONING
FOR MINIMISING LONG-TERM ENVIRONMENTAL PROBLEMS
IN TAILINGS DISPOSAL

RÉUNION DE TRAVAIL SUR

LES PROCEDES DE TRAITEMENT DU MINERAI D'URANIUM
ET DE CONDITIONNEMENT DES RESIDUS SUSCEPTIBLES D'ATTENUER
LES PROBLEMES D'ENVIRONNEMENT A LONG TERME

TABLE OF CONTENTS
TABLE DES MATIERES

SUMMARY OF THE WORKSHOP

D.M. Taylor (NEA) 95

PHYSICO-CHEMICAL PROCESSES IN URANIUM MILL TAILINGS AND THEIR
RELATIONSHIP TO CONTAMINATION

G. Markos, K.J. Bush (United States) 99

THE PRECONCENTRATION OF URANIUM ORES

H.A. Simonsen (South Africa) 115

INVESTIGATION OF NITRIC ACID FOR REMOVAL OF NOXIOUS
RADIONUCLIDES FROM URANIUM ORE OR MILL TAILINGS

A.D. Ryon, W.D. Bond, F.J. Hurst, F.M. Scheitlin, F.G. Seeley
(United States) 139

CHLORIDE METALLURGY FOR URANIUM RECOVERY : CONCEPT AND COSTS

M.C. Campbell, G.M. Ritcey, E.G. Joe (Canada) 149

REDUCING THE ENVIRONMENTAL IMPACT OF URANIUM TAILINGS BY
PHYSICAL SEGREGATION AND SEPARATE DISPOSAL OF POTENTIALLY
HAZARDOUS FRACTIONS

D.M. Levins, R.J. Ring, G.A. Dunlop (Australia) 159

REDUCTION OF RADIONUCLIDE LEVELS IN URANIUM MINE TAILINGS

M.H.I. Baird, S. Banerjee, A. Corsini, D. Keller, S. Muthuswami,
I. Nirdosh, A. Nixon, A. Pidruczny, M. Tsezos, S. Vijayan,
D.R. Woods (Canada) 173

A METHOD OF COMPUTING THE RATE OF LEACHING OF RADIONUCLIDES
FROM ABANDONED URANIUM MINE TAILINGS

C.J. Bland (Canada) 195

PRELIMINARY EVALUATION OF URANIUM MILL TAILINGS CONDITIONING
AS AN ALTERNATIVE REMEDIAL ACTION TECHNOLOGY
D.R. Dreesen, E.J. Cokal, E.F. Thode, L.E. Wangen, J.M. Williams
(United States) 201

URANIUM TAILING DISPOSAL BY THE THICKENED TAILING DISCHARGE
SYSTEM
E.I. Robinsky (Canada) 215

LIST OF PARTICIPANTS - LISTE DES PARTICIPANTS 229

SUMMARY OF THE WORKSHOP

Derek M. TAYLOR
Nuclear Development Division
OECD Nuclear Energy Agency

ABSTRACT

The Workshop on uranium ore processing/tailings conditioning
reviewed the various technologies by which the levels of pollutants
in mill tailings may be reduced. The participants agreed that the
alternative technologies could be presented as three options : discharge of tailings at a higher solids content than presently used ;
physical segregation of the ore or tailings into fractions, using
well known mineral dressing technology ; and concentration and isolation of the contaminants using more aggressive leaching processes. It
was emphasized that : the technologies of ore processing and tailings
conditioning are very site specific ; that radium may not always be
the major cause for concern in tailings ; and whichever technology
was followed, it would almost certainly be cheaper than any future
remedial action. It was agreed that work should continue in support
of conventional technologies unless the need for alternative technologies could be clearly demonstrated.

INTRODUCTION

        The nature and properties of uranium mill tailings at the
time of their disposal have a major influence on the potential impact
the tailings might have in the long-term. It was therefore recognised
early in the NEA programme on the long-term aspects of the management
and disposal of uranium mill tailings (1) that it would be useful to
hold a workshop on uranium ore processing and tailings conditioning
methods that could help to minimise this impact.

        Early in 1981, the NEA/IAEA Working Group on Uranium Extrac-
tion offered to organise a workshop on behalf of the Co-ordinating
Group on the Management of Uranium Mill Tailings and its three Working
Groups. The Workshop was sponsored by the Nuclear Energy Agency and
the U.S. Department of Energy, and was held at Colorado State University, Fort Collins, Colorado, USA, on 28th-29th October 1981.

        The meeting was attended by 24 persons representing five
countries, the NEA and IAEA.

OBJECTIVES OF THE WORKSHOP

        The objectives of the Workshop were :

    a) to review and discuss the various technologies by which
       levels of pollutants in mill tailings could be reduced and,
       where possible, to indicate the costs of such processes ;

    b) to identify, if possible, changes that could be made to the
       nature and properties of the waste material which could
       improve its long-term stability ;

    c) to make recommendations, if judged necessary, for further
       work in the field.

PRESENTATION OF TECHNICAL PAPERS

        Following a review of the physico-chemical processes in
uranium mill tailings and their relationship to contamination (Markos
and Bush), there were five papers on alternative processing technologies (Simonson, Ryon et al., Campbell et al., Levins et al., and
Baird et al.), a paper describing a method of computing the rate of
leaching of radionuclides from abandoned tailings (Bland), a review
of the studies under way on the conditioning of tailings (Dreeson et
al.) and details of a thickened tailings disposal system (Robinsky).
These papers are reproduced in this section of the proceedings.

GENERAL DISCUSSION ON PRESENTATIONS

        In addition to the discussions following individual presenta-
tions there was a general discussion on the alternative technologies.
A major contribution to the discussion was made by *Mr. C. Lendrum*
(Canada) who presented a critique of the chlorine/chloride based
processes which suggested that their costs would significantly exceed
those of existing processes and would do little, if anything, to
improve the environmental impact of uranium milling. He recommended
that a major effort should be made to convince health authorities to
establish scientifically based limits for the pollutants. This would
require the collection of basic data on which to base environmental
standards and regulations but could avoid anomalies such as the one
in Canada where the environmental guidelines for effluent from mills

(1) The background and terms of reference of this programme have been
    described by P.J. Rafferty earlier in these proceedings.

ask for levels of radium to be nearly an order of magnitude less than the Department of Health has set for drinking water. He also recommended that before large amounts of money are spent on research on new processes a preliminary evaluation should be made on the effects of the processes on the environment and stressed the importance of improving current technology.

Mr. E. Landa (United States) described the results of some leaching experiments on uranium ore and mill tailings carried out by the US Geological Survey using distilled water, sodium chloride solutions, acetic acid, hydroxylamine with hydrochloric acid, and alkaline DTPA. The tests showed that radium, thorium and uranium were more water soluble in the tailings than in the ore, that the uranium was in an easily leachable form in the two media and that radium was most effectively leached by the hydroxylamine. Other studies had shown that radium was moving in the tailings at the rate of about one metre per year.

In the remainder of the discussion it became clear that the participants were not convinced that mill tailings released under present technology presented an environmental hazard. However, if there really was a problem then an acceptable solution in terms of permissible levels of radionuclides should be defined and agreed so that the ore processing metallurgists would not have *to deal with a moving target*. Several participants raised the point that non-radiological contaminants probably presented a greater problem than radiological ones and recommended close liaison with the base metal industry at all stages of later studies.

SUMMARY AND CONCLUSIONS

The conclusions of the Workshop were as follows :

1. The participants agreed that the alternative technologies discussed could be presented as three options, as follows :

Option 1

Tailings could be discharged at a higher solids content than is presently used. It was possible that thickened discharge, at 65 % solids, could eliminate much of the seepage which was usually the most damaging aspect of tailings disposal. Other advantages of thickened discharge were that evaporation could be increased and dusting could be minimised. In addition, thickened discharge could be achieved using current technology. The initial and operating costs could be generally lower than present practice especially as the construction of high and costly dykes and dams could be eliminated.

Option 2

This option was based on physical segregation of the ore or tailings into fractions, using well known mineral dressing technology for separate treatment and disposal.

Several methods are in use for the preconcentration of uranium ores as these could have important economic benefits. In some instances the low uranium fraction of the ore, which was not to be treated further, could be disposed of without any conditioning while the remaining ore has a smaller volume and results in a smaller volume of tailings.

Studies have shown that radionuclides in tailings are distributed unevenly between sands (> 75 microns) and slimes (< 75 microns) fractions with a distinct preference for the latter. Splitting the sand and slimes could result in radionuclide-poor and radionuclide-rich fractions which could then be dealt with separately.

In general, the tailings would consist of three fractions pyrite, sands and slimes which could be disposed of separately.

i)   *Pyrite fraction* - In some areas of the world this fraction could be used to produce the necessary sulphuric acid for the leaching process. In other areas it could be buried away from the other tailings, preferably in an alkaline environment.

ii)   *Sand fraction* - If the ore was mined underground then this fraction could be used as backfill. Where open-cut mining was used this low-radionuclide fraction could be stored in a conventional tailings dam or returned to the pit.

iii)   *Slimes fraction* - The slimes could be stored alone in a tailings dam but a more satisfactory approach for some slimes would be to dewater them (on a belt filter), granulate them (either with or without a cement binder), and possibly to fire them (thermal stabilisation).

The costs of producing the different fractions was in the order of $ 1 per tonne of material. The cost of pelletizing the slimes would add between 50 cents and $ 1 to the cost depending on whether a binder was or was not used. The cost of firing the slimes would increase the cost by between $ 10 and $ 20 per tonne of slimes depending on the location and the process used (in particular the temperature required). However as the volume of the slimes was often relatively small compared with the total tailings even the cost of the most expensive option was not prohibitive.

Option 3

This option depended on concentration and isolation of the contaminants. A more aggressive leaching would be used to mobilise the contaminants. Nitric acid, hydrochloric acid and gaseous chlorine have been used experimentally to dissolve contaminants in solution and later remove them by ion exchange or solvent extraction. The resulting tailings may contain as little as 20 pCi/g of radium.

These methods have only been studied at the bench scale and it would be at least a decade before they could be used on the industrial scale. In addition, several problems were still to be resolved such as removal of chlorides and nitrates from the tailings and the disposal of the relatively large amounts of barium sulphate/radium precipitate.

The costs of these methods were difficult to estimate but would be in the region of $ 5 to $ 10/lb U above the cost of a sulphuric acid leach.

2.   The participants in the Workshop wished to point out :

a) that the technologies discussed are not always applicable in every case, i.e. ore processing and tailings conditioning are very site specific ;

b) that radium may not always be the major cause for concern in the tailings, and

c) whichever technology was finally followed, it would almost certainly be cheaper than any future remedial action.

3.   The participants agreed that before any recommendation could be made for further research on the subject, reactions to the discussions at the Workshop would be required from the Co-ordinating Group on the Management of Uranium Mill Tailings. Until this feed-back was received, work should continue in support of conventional technology rather than extending work on alternative new technologies.

PHYSICO-CHEMICAL PROCESSES IN URANIUM MILL TAILINGS AND THEIR RELATIONSHIP TO CONTAMINATION

G. Markos and K.J. Bush
GEC Research, Inc.
Rapid City, South Dakota, U.S.A.

ABSTRACT

Investigations over the last three years, of the physico-chemical properties of abandoned uranium mill tailings, in a range of climatic and geologic
environments of the U.S.A. show that tailings are in chemical disequilibrium
and are reactive due to their high salt and moisture content. The chemical
reactions redistribute the water and salts causing physical forces to operate
within the tailings which are manifested by many observable features. The
chemical reactions are the ultimate driving force causing chemical and physical
instability and must be considered in the development of safe long-term
disposal practices. Chemical reactions are also significant at the contact of
two differing chemistries of the tailings and environment where neutralization,
precipitation, and immobilization of contaminants may occur.

1.   INTRODUCTION

Uranium recovery throughout this century has resulted in many million tons
of tailings laying unconfined on the land surface. Current uranium recovery
activities are adding to this mass of waste. As these wastes contain large
quantities of toxic components, disposal practices and management procedures
with long term containment potential are required. The matter of concern
concentrates primarily on the radionuclide toxicity. Chemically toxic components such as cadmium, arsenic, selenium, chromium, lead, and others, need
consideration because their increased presence in the environment may create a
health hazard. An understanding of the chemical properties of these wastes and
their interaction with their environment is needed in order to develop economic
and stable containment technologies.

A complex geochemical system exists between the chemical, physical, and
hydrological properties of both the tailings pile and the environment. Physico-
chemical and kinetic processes superimposed on the transporting medium of water
govern the migration of the toxic components. The containment of toxic components must, therefore, be evaluated in terms of the chemical properties of
the components along with the influence of the chemical parameters of the
environment which mobilize or retard the movement of soluble substances. In
addition to the consideration of the chemical parameters, the physical stability of the bulk solids are included in the evaluation. The high amounts of
water soluble salts and the chemical disequilibrium conditions have a profound
effect on the physical properties, chemical reactivities, and migration of
contaminants as well as on the prevailing hydrologic nature of the wastes.

The safe long-term disposal and management procedures of wastes require
addressing several questions:

(a) Are tailings physically stable so that engineered barriers of synthetic or
     natural materials intended to confine toxic substances will remain for a
     desired period of time?

(b) Do relationships exist between physical stability and chemical reactions
     in the tailings, and if so, how are the physical properties affected by
     the intrinsic chemical properties of the tailings and the changes caused
     by chemical disequilibria?

(c) What factors control the chemical mobility of components in wastes which
     will allow the undesired substances to be transported by water into the
     human habitat?

(d) What chemical interaction exists between the pore waters in the waste
     materials and those in the natural material upon which the wastes are em-
     placed, and, is it possible to utilize or modify these interactions by a
     relatively simple chemical means to prevent seepage of contaminants into
     the shallow subsurface water?

(e) What alternatives to conventional liner and cover technology could be
     introduced in order to dispose of wastes economically and safely for a
     long period of time?

For these questions, it is difficult to provide satisfactory answers now
or in the immediate future because of a lack of adequate understanding of the
properties of mine and mill wastes. However, current work has illustrated the
gross characteristics of the geochemical system and the interrelated complexi-
ties between the chemical, physical, and hydrological components of the system.

This paper constitutes a brief summary of the geochemical investigations
of uranium mill wastes which is currently in progress, sponsored by the Uranium
Mill Tailings Remedial Action Program of the U.S. Department of Energy. The
enumerated questions above are based on information from the investigation.
This paper attempts to illustrate the basis for the questions and to elucidate
some of the properties and characteristics of the system in view of those
questions. Physical manifestations of chemical changes, induced by the disequilibrium of tailings materials, and the mobility of toxic substances at

interfaces, constitute the main theme of this paper. Based on these considerations some recommendations for management practices are also included.

2.   FUNDAMENTAL CONCEPTS CONCERNING TAILINGS IN THE ENVIRONMENT

Variation in physical configuration and in the chemical properties exists
between tailings piles and within a single pile. Variation within a pile may
often be accounted for by changes in milling and waste management methodology
during the active life time of the tailings. Parameters responsible for
variability in tailings are mineralogy of feed ore, methodology in extractive
metallurgy, engineering practice of tailings emplacement, and post operational
treatment. Superimposed on the conditions of the tailings materials are the
conditions of the environment such as topographic, hydrographic, and climatic
conditions which vary locally. Consequently, a complex geochemical system is
developed between the variable properties intrinsic to the tailings pile and
the variable properties of its surrounding which is responsible for the
interactions between the tailings and the environment.

Uranium mill tailings contain chemically and radiologically toxic materials in high concentrations. These substances may move from the tailings into
the human habitat and cause harmful effects. The main pathways of movements
are by solution and suspension with moving fluids (water and air) and by
physical and chemical transfers as diffusion, surface tension, dissolution, and
sorption. Evaluating potential or actual contamination from the tailings
necessitates the study of the geochemical relationships between the contaminants and the tailings environment to determine the potential for mobilization
of the contaminants and the potential for movement.

Uranium mill tailings are residues from uranium ores removed from a
subsurface geochemical environment and subjected to a vigorous chemical and
physical processing to remove the ore values. Removal from its subsurface
environment changes the chemical parameters and moisture conditions of the
original ore. Roasting, chemical leaching and oxidation change the chemical
relationships among components and redistributes the soluble materials. All of
these changes result in a chemical disequilibrium. It follows from Le
Chatelier's principle that a readjustment of the perturbed system must take
place to establish a new equilibrium compatible with the new environment. As a
consequence, tailings are chemically dynamic bodies where chemical reactions
are in progress to reach some new equilibrium or steady state conditions.

The nature of the dynamic processes between the tailings and the environment is influenced by the conditions of the surroundings. Particularly
important are the proximity and distribution of water available to the tailings
pile that are determined by the local topographic and geomorphologic conditions. Drilling and sampling of the tailings, [1], [2], and [3], have shown
that the parts of tailings piles resting on gullies or buried channels contain
more pore water than other parts and where a shallow lying water table
intercept the tailings a water dome may form.

Interaction between the tailings pile and its surroundings also relates to
the physical and chemical properties of the underlying soil and bedrock. Soil,
bedrock, and clay underlayment used as liners are porous materials. Particle
size distributions, bulk density measurements, and percolation tests on small
core samples are the methods commonly used to determine the hydraulic conductivity. However, these methods fail to measure the fracture porosity which tends
to exist in any natural material (space between soil structural units, local
desiccation due to the presence of salts and loading cause fractures in soils
and bedrock, including shales) which should be considered as pathways for the
flow of water.

3.   INTRINSIC PROPERTIES OF TAILINGS

3.1 Properties of salts in tailings

Large quantities of salts and their heterogeneous distribution play the
most significant role in controlling the dynamic characteristics of tailings.
Many of the salts possess hygroscopic or deliquescent properties which allow

them to retain water as well as obtain it from various sources, either internal or external to the tailings. In addition, many of the hydrated salts (epsomite, $MgSO_4 \cdot 7H_2O$, for example) act as source or sink for water.

Moisture conditions of tailings that are salt controlled have been observed on many occasions. Only a few centimeters below the surface, tailings contain considerably higher water content than the surrounding soils, and often contain water saturated pore space. Saturated conditions can exist on the surface of the tailings with very high salt resulting in a "mud lake". In some extreme cases, the "mud lakes" remain on the tailings in the summer months in desert environments where daytime air temperatures exceed $40^{\circ}C$ accompanied by very low relative humidity and desiccating winds. These "mud lakes" may have a few millimeter to a few centimeter crust during the daytime, but by morning, a wet surface forms due to the hygroscopic salts. Salts obtain water by desiccating less saline materials which results in a crack pattern. Obtaining water, particularly from subsurface environments, may be accompanied by osmotic pressure inducing a physical flow of water.

Salts may contribute up to approximately twenty percent of the total weight of the solid matrix. Dissolution, redistribution, and reprecipitation cause large volume changes and may result in cavity development at one place and swelling at another place. Cavities of over a cubic meter or several meter depths following more than a hundred meter linear trend of a series of cavities have been observed. Salts in the tailings also accelerate the chemical destruction of various metals, organic materials and silicate minerals. High ionic strength of interstitial solutions imparted by the easily soluble salts and variations in pH and redox potential play a substantial role. Silicate dissolution has been described elsewhere [4], whereas effects on organic materials are given below.

3.2 Relationships between organic materials and salts

Apparently, an important set of interactions exist between organic materials and the salts of the tailings. Observation has verified: adsorption on cellulose base materials; dehydration of asphalt-water emulsion followed by adsorption and precipitation in the porous asphalt; complexing and mobilization of metals by soluble organic material.

Desiccation and destruction of an asphalt emulsion cover on the Grand Junction, Colorado tailings have been observed. Six weeks after the emplacement of asphalt, salts had moved through the 15 centimeter thick experimental asphalt cover and accumulated on the surface. Simultaneously, with the salt movement, the asphalt appeared porous and cracked. Chemical analysis of samples of asphalt and tailings below the cover indicated an order of magnitude increase of salts accompanied by heavy metal accumulation and an order of magnitude increase in gamma radiation in contrast to the uncovered adjacent areas. Slime areas with salt precipitates in the Grand Junction tailings show ten to fifty $\mu R/hr$ compared to 400 to 600 $\mu R/hr$ gross gamma and gross beta radiation on the asphalt. Nine months later, in many extensive areas, no traces of the fifteen centimeter thick asphalt cover was found. Similar deterioration of an experimental cover on Tuba City, Arizona tailings occurred a few years earlier. Here, the U.S. Bureau of Mines applied synthetic polymer admixed cover on the surface of the tailings [5]. At the time of our investigation of the tailings, twelve years after the emplacement of the cover, no signs of the cover were observable.

Chelation of metals by organic material may be a mechanism for the accumulation of salts. Chemical reaction with the salts such as redox reactions, hydrolysis, and hydration are speculated to initiate the destruction of organic materials. Heavy metals may also become significantly mobilized by chelation and upon seepage from the tailings, significant migration into the external environment may occur.

3.3 Migration of salts to the surface of tailings

Salt precipitation on surfaces of tailings (with and without cover) has been observed on all of the twenty tailings investigated. Salts tend to migrate to the surface of the interface with the atmosphere where they precipitate. These salts carry contaminants to the surface where they become susceptible to wind or water erosion.

Salts and trace metals are transported by water moving upward by capillary, thermal, chemical, and osmotic gradients. The presence of salts influence the gradients by variation in surface tension and the colligative properties of the solution, and pore size changes due to precipitation and dissolution reactions. Osmotic potentials and pressures may be a significant driving force for water redistribution and, hence, may influence the movement of salts.

With our current understanding of salt migration in tailings, it seems reasonable to conclude that none of the currently employed or experimentally emplaced cover materials will serve as a barrier to salt migration. Salt migration will ultimately cause destruction of the containment properties of the cover material which will allow exposure and release of the toxic materials.

3.4 Physical effects attributed to chemical reactions

Chemical reactions in tailings cause many physical effects which are directly observable as surface features on the tailings piles. The magnitude and duration of these features are not well established; however, the observable manifestations have occurred since the cessation of operations about fifteen to twenty-five years ago. Many of the features, however, formed during the time of this investigation ranging from a few weeks to over a year period of time.

A detailed account of these features is impossible to provide within the framework of this communication, therefore, only a brief listing is provided. The features described here constitute a composite of the twenty tailings investigated so far. Not all the described features are present in all tailings and the magnitudes and distributions show considerable variations. The distribution may appear irregular or with a discernable trend. Linear trend patterns show some association with buried subsurface channels for water flow. Relationships of occurrences, however, are difficult to establish on the basis of existing data.

According to their physical appearance, the following types of physical features were established.

(a) Geometric crack patterns and associated mounds appear commonly in many tailings. The patterns can by polygonal or orthogonal. Mounds may form between crack delineated areas. These mounds range from a few centimeters to a few tens of centimeters in height and from about twenty-five centimeters to over a meter in diameter. Salts show segregation and preferential precipitation relative to cracks and mounds. The cracks, in most cases, are filled with loose material. Depth of the cracks show variation from a few centimeters to an undetermined value of over a meter. These prominent geometric patterns are associated with areas of high salt content in uncovered tailings or where salts prominently move through cover material. The possible explanation for development of geometric crack patterns and associated mounds is desiccation and evaporation on the surfaces by thermal gradients.

(b) Cavity development and "piping" or tunneling are common features on many tailings. Tunnels are present in vertical, horizontal, or oblique direction. Their size ranges from a few centimeters to several tens of centimeters in diameter and up to several meters in length. In some cases the tunneling features are associated with dikes. Destruction of gravel dike by extensive gullying resulting from tunneling has occurred within a period of a year as were observed on several occasions. Cavities forming within a few weeks to a few months of time were also observed. The experimental asphalt cover collapsed after less than six months of emplacement, forming a cavity of over a meter diameter and approximately eighty centimeters deep. With time, the size of the cavity increased. Many other cavities are present on tailings. Some of them extend several meters below the surface. Several formed between two observation periods. The origin of these cavities are not well understood. A possible explanation includes salt and water redistribution processes. It is also possible that the surface charge changes due to reactions which cause flocculation or dispersal of colloidal size material promoting movement of suspended particles. Also, dissolution of the silicate matrix could cause the redistribution of large volumes of materials.

(c) Mounds of varying sizes are frequent in many tailings. Some of them are loose aggregates of materials similar to quicksand conditions. Others are associated with gas bubbles or pore pressure induced mounds. Higher than atmospheric pore pressures have been observed during drilling operations and excavation of pits. These pressures may contribute significantly to the development of mounds.

(d) Gas bubbles and mounds are another set of features found on tailings. Hydrogen sulfide gas in one tailings has been detected by its odor. Other gases, possibly carbon dioxide and hydrogen, may also form. During drilling operations the equipment was plated with copper accompanied with development of gases. We suspect this to be the reduction of the cupric ion in copper sulfate salt and the gas being an oxidation product. Mounds attributed to gases varies from a few millimeters to over eighty centimeters in diameter and occur in slime areas where salt concentrations are high.

(e) Other features observed include precipitation nodules over thirty millimeters in diameter, melting of snow and ice over salts on the surface of the tailings, and decay of vegetation planted on cover materials on the tailings.

3.5 Significance of intrinsic properties

The unevenly distributed high salt content of tailings constitutes the major aspects of disequilibria. The salts impart chemical reactions and redistribution of materials within the tailings which, in turn, result in physical instability through time. Many of the manifestations of physical instability indicate rapid changes and forecast a problem on current practices of containment.

Aside from observations and inferences about the cause-and-effect relationships between chemical reactivity and physical activity of tailings, the physico-chemical processes are little understood. For a time-lasting and safe rehabilitation of tailings the physico-chemical relationships cannot be ignored. Research in details of cause-and-effect relationships are necessary to alleviate the physical instabilities and to provide "real world" data for engineering design of containment practices.

Although intrinsic properties of tailings and relationships to the surrounding environment are site-specific conditions, using physical chemistry to evaluate cause-and-effect relationships can provide a common denominator to all tailings: energetics. With site-specific data available predictions on tailings behavior will be possible when the energetics of chemical and physical relationships become established. Until then, short range experiments seem futile as experience shows.

4. TAILINGS-ENVIRONMENT GEOCHEMICAL SYSTEM

In the tailings-surroundings system, three parts exist: the tailings, the adjoining surroundings, and the interface region between the tailings and surroundings. The interface is a distinct region with properties characteristic of the mixing of the different chemical and physical properties of the tailings and surroundings. A model for the chemistry of the system which predicts the interactions between the tailings and the soil must be particularly cognitive to the properties observed in the interface region.

The necessary first step in coming to an understanding about the chemistry of the interface requires that characterization be made of the geochemical system within the tailings and the surroundings and of the hydrological properties of the transporting medium. The components of the chemical system, the reactions involving the components, and the parameters controlling soluble and insoluble phases must be determined.

The tailings system is heterogeneous and complex. Data analysis shows a system which is dominated by the following characteristics:

(a)   high ionic strength in interstitial pore water solutions,
(b)   easily soluble salts of calcium and magnesium sulfates and sodium chloride,
(c)   electroactive species of uranium, iron, manganese, and vanadium,
(d)   low pH values relative to the surrounding environment.

One of the most significant aspects of the tailings-soil-water system is the large chemical differences between the pore solution in the tailings and pore solution in the subjacent soil. When the chemically different pore solutions contact each other and mix at the boundary chemical reactions take place. An interface zone is formed which has different properties and composition from those of solutions in the adjacent materials.

Variations in the chemistry of the environment in which a solution moves through will cause chemical reactions in an attempt to establish equilibrium with the new conditions. Both the solid matrix and the solution may acquire significantly different properties as the reactions take place. In the tailings-soil-water system these reactions result in creating sources and sinks for the various components.

Reactions due to interaction at interfaces are readily observed on core samples of and in test pits in tailings. Bands of precipitated iron, manganese, and uranium, along with gypsum and other salts are often present at the contacts between tailings and subjacent soil, between tailings and cover material, and between parts of tailings where slime is in contact with sand. These bands were observed in both water saturated areas and in drier areas where evaporation or desiccation has concentrated the solutes. Chemical analysis and limited X-ray diffraction analysis has confirmed the visual observations.

X-ray diffraction analysis have indicated the precipitation of amorphous ferric hydroxide, jarosite $(KFe_3(SO_4)_2(OH)_6)$, alunite $(KAl_3(SO_4)_2(OH)_6)$, amorphous silica and alumino-silicates, gypsum $(CaSO_4 \cdot 2H_2O)$, epsomite $(MgSO_4 \cdot 7H_2O)$, anhydrite $(CaSO_4)$ and bassanite $(CaSO_4 \cdot 1/2H_2O)$. Other precipitates were also present but their positive identification presented problems because of solid solution formation, poor crystallization, their presence in small quantities, and lack of single-crystal analysis.

These interfaces represent significant boundaries where physical changes in addition to chemical reactions can also take place. Precipitation in a sparingly soluble form reduces the size of effective pore space and may strongly modify hydrodynamic conditions. Precipitation of colloidal particles may drastically alter the electrochemical properties of the surface of the solid matrix. Silicates are cation adsorbers because the silicate mineral assemblage provides a wide range of negative surfaces. Sesquioxides and oxyhydroxides of iron, aluminum and manganese by their positively charged surfaces may be important anion adsorbers. The precipitation of a coating of iron hydroxide on a silicate surface has the potential to change the adsorptive characteristics of the solid matrix. This change may tend to release the cations originally adsorbed. The precipitation of the colloidal particles could also effectively form a perm-selective membrane. The concentration differences between the two sides of this perm-selective membrane may cause the movement of solvent from the dilute to the more concentrated side of this perm-selective membrane through the formation of osmotic potential, thus an osmotic pressure. This pressure, if it exists, reverses the expected direction of water movement.

Two major sink conditions are formed at the interface: (1) precipitation because of supersaturation and/or (2) removal by a host through coprecipitation, occlusion or adsorption. Precipitation is a spontaneous process in supersaturated solutions. For nucleation of precipitate formation an energy barrier must be surmounted to form a new phase from the solution [6]. In a heterogeneous nucleation process, which is the predominant crystal forming process in geochemical systems favorable lattice patterns are available on existing solid surfaces and precipitation can be a rapid process.

Supersaturation can occur for many reasons. One reason is mixing of two or more solvents having different solute compositions. Runnells [7] has shown that mixing of two solutions which are both undersaturated with respect to a given mineral may produce a mixture which is either supersaturated or more pronouncedly undersaturated depending on the chemical properties of the original solutions and on the mineral in question. Mixing of waters of widely differing chemical compositions may easily result in supersaturated conditions [8].

The high concentrations of electroactive components, their precipitation behavior at lower Eh and higher pH values, and their known scavanging capacity comes to the focus of attention in evaluating the interface processes. Precipitation of silica, alumina, and alumino silicate gel form another set of scavanging substances for trace metals. Soluble aluminum and silica are present in the tailings in high concentrations. This is attributed to the destruction of silicates at the low pH high sulfate conditions of the tailings. Destruction of silicate materials in the extraction process of uranium has been established by Merritt [9] and continued destruction of the silicates in the tailings has been recognized by X-ray diffraction analysis and chemical analysis by Bush and Markos [10].

5.   GENERAL Eh-pH RELATIONSHIPS

5.1 Introduction

Measurements of Eh and pH provide a sound basis for the qualitative understanding and prediction of the chemical interrelationships in the tailings-soil-water system. Investigation of the interactions between the tailings and the soils based upon Eh and pH measurements of interstitial solutions has been done by three methods: (1) Eh-pH diagrams representing the conditions of the tailings and of the soils; (2) vertical distributions of pH and Eh values through the tailings into the soil (3) mineralogic analysis on samples of tailings and subtailings soils at the interface.

5.2 Eh-pH diagrams

On the basis of the statistical treatment of data available for several tailings piles large variations in both Eh and pH values exist within a single tailings and among different tailings. The statistical mean values and the values of mean plus one standard deviation of Eh and pH together with regression lines for each data set, are shown in Figure 1A.

The distributions of Eh to pH are similar for the Riverton and Salt Lake City tailings but differ for the Grand Junction tailings. The Riverton and Salt Lake City tailings resemble each other in terms of both the mean values and distribution pattern of Eh and pH about the respective means. The spread of Eh values (for the units of mV used in the diagram) is larger than of pH for the former two similar tailings, whereas the spread of pH is greater than the spread of Eh in Grand Junction. The spread of Eh, however, is similar for all the three tailings.

The mean values Eh and pH in the Riverton and Salt Lake City tailings are close to each other but differ from those in the Grand Junction tailings, the mean Eh being about 100 millivolts lower and the mean pH about 2 units higher for the Grand Junction tailings. The variation between tailings can be explained on the basis of post-operational treatments. Although all of these tailings resulted from acid-leach processes, the Grand Junction tailings was pH-neutralized by the addition of ammonium carbonate [9].

Another important aspect the data, represented by Figure 1, is the very large range of both Eh and pH values within a single tailings pile. This fact raises important questions: First, what kind of interpretation is possible regarding the tailings as a whole? Second, what approach is required to evaluate the details of the prevailing chemical conditions of the tailings-soil-water system? In the evaluation of the interactions between the tailings and the soil, the tailings can be treated as a single unit despite the large spread in the Eh and pH values since they are significantly different from the soils. However, variation in the chemical system within a tailings and between tailings is too great to allow the details of the tailings to be evaluated as one group. Evaluation of the details controlling the mobility of the components within the tailings requires classification of the data into groups exhibiting similar chemical behavior [11].

Tailings-soil-water systems based upon Eh and pH measurements are represented for three sites: Grand Junction, Salt Lake City, and Riverton, shown in Figures 1B, 1C, and 1D respectively. The diagrams illustrate the differences in the conditions between the tailings, the subtailings soil, and the background soils.

The general differences between the arithmetic means of the pH values of subtailings soil and of the background soil are similar for all three sites, but the directions and magnitudes of the differences between the data from the two types of soil vary for each site. In Riverton and Grand Junction, the subtailings soils have a lower pH than the pH of background soils. This is in contrast to the relationship in Salt Lake City where the pH is higher in the subtailings soil than in the background soils. The direction of difference in

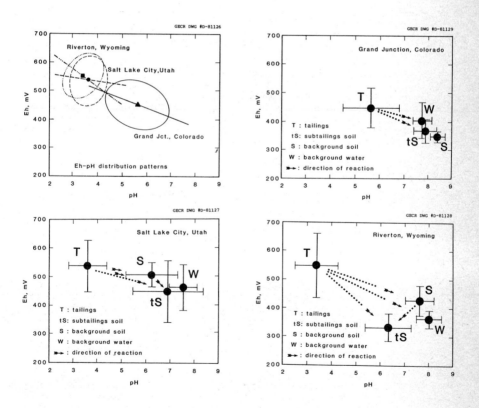

Figure 1. Relationships of tailings and their environments as they relate to conditions of Eh and pH. Mean values with one standard deviation are represented for three tailings.

the Eh values between the subtailing soil and background soil is the same in Riverton and Salt Lake City, while the soils in Grand Junction exhibit a pattern of Eh variation in the opposing direction.

The variations in the pattern of behavior of the Eh and pH in the soils from the three sites may be due to the differences in the soil types and in the hydrologic conditions of the area which influence the alterations within the soils buried under the tailings. The soils at Riverton and Grand Junction have formed on overbank deposits of larger rivers deriving their sediment load from crystalline and sedimentary rocks in mountain areas. The soil in Salt Lake City area has formed in a lake basin on an interfingered deposit of saline lake deposits and clastics derived from sedimentary and crystalline rocks of nearby mountain areas. The soils in Salt Lake City contain stagnant waters, rich in organic materials and in many places are saturated nearly up to the surface. The soils in Riverton and Grand Junction are better drained although a shallow water table prevails.

This descriptive model has no capability to resolve the details of the processes that have taken place. However, the processes that occur and the changes induced are important considerations in evaluating the long term results of tailing management and they warrant further investigation.

5.3 Vertical distribution of Eh and pH conditions

The distribution of pH and Eh values through the tailings into the subjacent soil expresses two important properties:

(a) The pH and Eh values show changes in opposing directions. This type of change is governed by redox conditions associated with hydrolysis re-actions. A typical example of this type of reaction is the precipitation and dissolution of ferric hydroxide.

(b) The second important property which the diagram expresses is the abrupt change in the pH and Eh conditions across the interface at the base of tailings. The pH increases approximately four units within a distance of a few centimeters. Within this same interval the Eh decreases by about 150 millivolts. There are variations in the magnitudes and abruptness of changes across the interface but all locations investigated show this general pattern.

The abrupt change in pH and Eh at the base of the tailings leads to three possible interpretations of the mechanism:

(a) The solution from the tailings does not move into the subtailings water system.

(b) Solution from the tailings may move into the subjacent soil, but the water in the soil is so well buffered with respect to pH and poised with respect to Eh that the mixing effects of the tailings water with the soil water is negligible or undetectable.

(c) The rate of movement or seepage of water from the tailings to the soil water is negligibly low compared to the rate of replacement by uncontaminat-ed ground or soil water and dilution becomes a sufficient mechanism to eliminate the effect of acid. A pH change of four units, however, require a 1:10,000 dilution.

Although construction of vertical distribution diagrams illustrate the qualitative properties of materials, determination of the mechanism requires a detailed investigation. A quantitative evaluation could be expressed by determination of buffer capacities and redox poising capacities of the solu-tions in question. The best method is by acid-base and coulometric titration respectively. This type of work was not included in this project and no data of this type is known to exist from tailings.

5.4 Mineralogical analysis of acid penetration below the tailings

A case study was made for Salt Lake City tailings to investigate the effectiveness of acid and sulfate penetration into the natural ground through the interface [10]. One location selected for this study was the slime area of tailings where ponds of water have existed and where a constant hydraulic head was maintained for long periods of time which would maximize the potential for seepage below the tailings. Another area selected was the sandy part of the tailings in which a more intensive water movement from the tailings into the ground might be expected. The investigation was carried out using X-ray diffraction techniques to detect mineralogic changes relating to acid-soluble minerals.

Sulfuric acid was a major constituent of the primary hydrometallurgical process to remove uranium from the ore. As a result the tailings contain high concentrations of sulfate ion--up to 9.3 weight percent in tailings, and as much as 69 percent of separated water-soluble salts, resulting in pH values of about 3. Sulfate ion is expected to be mobile in most environments because of its negative charge. Hydrogen ion is expected to be mobile due to dissociation equilibria with some oxyanions and other weak acids.

Calcite is a common mineral in soils of the Salt Lake City Valley which include the subtailings soil. Calcite is the dominant mineral throughout the 120 cm depth investigated with the presence of dolomite and a small amount of plagioclase in the upper part of the profile. Gypsum ($CaSO_4 \cdot 2H_2O$) is absent in the soils investigated.

Calcite is soluble in cold acid solutions. Consequently, if acids have penetrated into the subtailings soil, loss of calcite by dissolution should occur. In acidic solutions containing sulfate, calcite dissolves and the calcium reprecipitates as gypsum, as is well-known from studies of diagenetic alteration of sediments. In less acidic solutions the surface of calcite is replaced by gypsum which forms a protective cover preventing further destruction of calcite.

X-ray diffraction studies of the mineralogy across the interface are summarized in Figure 2. Within the tailings, gypsum is the dominant calcium-bearing mineral which was formed after the ore processing. The acid added in the milling dissolved calcite present in the ore, and the available sulfate caused precipitation of gypsum according to the reaction

$$CaCO_3 + SO_4^{2-} + 2H^+ + H_2O = CaSO_4 \cdot 2H_2O + CO_2 \uparrow \qquad (1)$$

Below the base of the tailings, in a five centimeter thick upper part of the subtailings soil, calcite coexists with minor amounts of gypsum. This suggests that some sulfuric acid has penetrated from the tailings and caused replacement of some of the calcite with gypsum. The soil is apparently well-buffered, and has neutralized the acid, as is suggested by the sharp pH change from 4.9 to 7.4 within a distance of nine centimeters. The pH of the water extracts of the soils below the tailings is maintained between 7.4 and 7.9. The same conditions were found at the second location investigated.

The mineralogy of the soil 39 cm below the base of the tailings shows the presence of calcite and a lack of any trace of gypsum. It is apparent that the effective penetration of acid into the subtailings soil is less than 39 cm at this location. The preservation of a high pH in the subtailings soils could be explained by a high buffering capacity of the soil, but the mineralogical analysis indicate a lack of movement below the interface zone.

Mineralogic evidences indicate that neither acid nor sulfate ion have migrated into the water saturated pore space of subtailings soils in quantities characteristic for the tailings solution. It can be assumed, if one of the most mobile ions does not penetrate into the soil the more immobile ions will be retained. This assumption, however, should be verified by other tests such as indicated in the preceeding section.

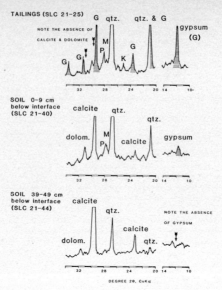

Figure 2.   Mineralogy of tailings and subtailings soil through the interface.   X-ray diffraction analysis was made on core samples.

5.5   Results and interpretation

Figure 3 shows the predominance diagrams for iron, manganese, and uranium
which were calculated specifically for the typical conditions of the acid-
leached tailings in Riverton and Salt Lake City. The equilibrium lines for the
dissolution-precipitation reactions were calculated for the chemical conditions
within the tailings and for the chemical conditions outside of the tailings.
The equilibrium in the tailings is represented by a solid line and the
equilibrium outside the tailings is shown using a dashed line. The measured Eh
and pH conditions of the tailings and their respective environments are plotted
on the diagrams.

The diagrams reveal many important conditions.   Under tailings conditions
the solid species of manganese and uranium are thermodynamically unstable.
Iron, manganese, and uranium in the aqueous phase are most likely in the form
of sulfate complexes.   The complexed species are quite stable and allow high
concentration of iron and uranium to be soluble.   Some tailings solutions lie
in the thermodynamic stability fields of solid phases iron, manganese, and
uranium.   The positions of these points are in agreement with the data.   Some
tailings samples exhibit low concentrations of iron, manganese, and uranium in
the aqueous phase relative to the total uranium and iron.

The diagrams also suggest that upon dilution, and without the increase of
pH or decrease in Eh, the field of stable aqueous species become enlarged. This
leads to an important interpretation.   If tailings solutions move into a more
dilute and poorly buffered pH environment which will allow the acidic conditions of the tailings to prevail, iron, manganese, and uranium will remain in
solution.   The metals and associated contaminants will be susceptible to
transport if no significant adsorption takes place along the path of water.
However, if the migrating solution encounters high pH or reducing conditions,
precipitation of iron, manganese, and uranium should occur.   Such conditions
exist in most ground waters and organic rich soils.

Comparing   the   conditions   of   tailings   solutions   with   those   of   the

- 110 -

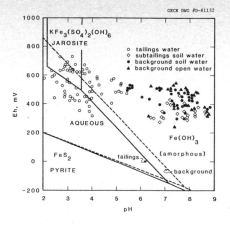

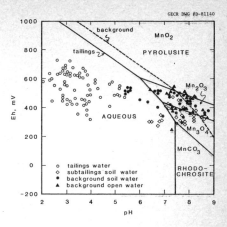

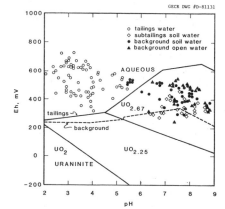

Basis for calculations:

	Tailings	Background
pSO_4^{2-}	1.3	2.5
$pHCO_3^-$	5.0	5.0
pFe_t	2.5	6.0
pMn_t	2.5	6.0
pU_t	3.0	6.0
pK^+	4.0	4.0

Where p designates the negative logarithm of activity.

Figure 3. Relationships of tailings and background solutions to thermo-dynamically predict precipitates of iron, manganese, and uranium as functions of Eh, pH, and activities of components. Calculations were made with respect to existing conditions in tailings sites.

surrounding environment suggests that iron and uranium should precipitate upon movement to an environment with an increased pH and a slightly decreased Eh. In well buffered and poised systems encountered in the investigations, the tail-ings solutions, if entered and mixed into the subjacent or adjacent waters would acquire the Eh and pH characteristics of that water. Consequently, the precipitation reactions should occur. Upon precipitation, iron and manganese, and perhaps uranium, act as important sinks for contaminants in actual or poten-tial transport by scavenging trace metals [12]. Evaluation of the movements of trace metals, therefore, necessitates the assessment of the precipitation behavior of iron, manganese, and uranium.

A significant point should be stressed here. Evaluating the relationships of tailings solutions to the waters in the surroundings and predicting the potential precipitation must include determination of the intensity factors, Eh and pH, the capacity factors, pH-buffering and Eh-poising, and the concentra-tions of the chemical components in the respective solutions from the tailings and background. This set of information can be used to calculate the changes that should occur in waters from the tailings if they mix with the waters in the surroundings.

Two examples of field verification of the Eh-pH diagram are shown below. The methods include the evaluation of vertical distributions of soluble contaminants across the interface [1] and statistical analysis of the relationship between the quantity of contaminants associated with the quantity of precipitated iron hydroxide [2]. In addition, a detailed study of the distribution of radionuclides into the soils from the tailings at Salt Lake City were completed by Markos and Bush [10] and show migration of less than one meter. Figure 4 illustrates the vertical distributions. Note the abrupt changes at the interfaces. In many cases an accumulation is present on the top of the interface region. These examples from the Salt Lake City tailings are representative of most tailings sites we have investigated.

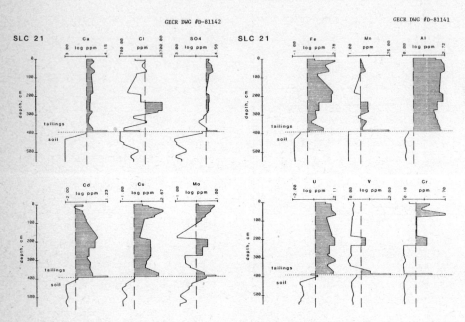

Figure 4. Typical vertical distribution of water leachable components of tailings and subtailings soil from core samples through the tailings into the subjacent soil.

Correlation analysis of trace metals with iron that was solubilized in acid extracts (HCl pH 2) from the Canonsburg tailings has been made. Good correlations exist between iron and trace metals such as arsenic, cobalt, chromium, nickel and uranium in the higher concentration ranges of the trace metals. Arsenic shows the statistically most explained relationship (R^2 = 0.87) followed by cobalt and chromium and nickel. Two regression lines were generated for the relationship between iron and uranium, suggesting a possibility of two different associations. No resolution for the differences can be offered, however.

The relationships do not explain the reasons or chemical mechanisms of association between iron and the other components. Precipitation of iron, the possibility of scavenging and the precipitation of the other metals in the same chemical environment agree with the general understanding about the geochemical behavior of these components [12].

6.   CONCLUSIONS

Behavior and properties of uranium mill wastes result from the interactions among the feed ore, the method of hydrometallurgical processing used, and the external environment. The ore has been subjected to crushing and vigorous chemical processes after removal from its subsurface habitat. During the hydrometallurgical processing dissolution of many minerals took place together with the dissolution of the uranium containing minerals. Disposed material included the raffinate solution together with the solid matrix, which was ponded on part of the tailings. Evaporation caused the concentration and finally the precipitation of easily soluble salts and various sparingly soluble oxides and hydroxides. As the tailings emplacement continued these salts became incorporated into the bulk of the tailings.

This mass of solution and solid has been disposed of on natural or prepared surfaces which have interacted with the atmosphere and shallow subsurface water. This disposed mass is heterogeneous in terms of physical and chemical properties. Slime and sand are interfingered due to settling in pond beach areas or to mixing of the tailings by mechanical disturbances of reshaping. Water content varies within the tailings relative to the position of the ponds disposal (usually enriched with slime material) and movements of water within the tailings.

Tailings show considerable physical instability through time which have been interpreted to be a result of chemical reactions, predominantly controlled by mobile and easily soluble salts. This instability may affect the stability of engineered barriers to prevent contaminant movement from the tailings, unless the barriers are designed to withstand the corrosion, desiccation and volume changes caused by the salts. An alternative to designing appropriate barriers is to remove the salts from the tailings. The main geochemical controls of reactions within the tailings are the quantity of sulfate and chloride salts (which are unevenly distributed), the availability of water from external and internal sources, pH and Eh, and the potential gradients existing because of the chemical heterogeneity of the tailings.

Interactions between the tailings and the soil results in precipitation reactions, particularly of iron. Precipitation reactions of iron depends on the pH and Eh conditions where the tailings and soil solutions mix. Where the soil water system is well buffered precipitation of iron is caused by the maintenance of higher pH of the soil water.

The scavenging effects of the precipitating iron appear to be major controls in the preventing migration of contaminants into the environment. Precipitation causes the decrease of pore size and hydraulic transmissivity of the porous material. At the tailings-subtailings soil interface other colloidal precipitates, in addition to the amorphous iron, such as silicates and silica gel seem to develop a protective barrier against movement of contaminants. The changes in the interface zone produced by precipitation may result in a semi-permeable membrane which may reverse the flow of water by an osmotic potential. The osmotic pressure due to the flow of water from the shallow subsurface into the tailings is postulated to have effects on the physical stability of the tailings pile.

The consequences of movement of water and solute within the tailings can be speculated upon to be related to the chemical and physical instability. Movement of water and the consequent chemical reaction will continue until a chemical and hydrological equilibrium is reached within the tailings and between the tailings and the surrounding environment. At the present state of understanding, the chemical processes and the cause-and-effect relationship between chemical reactions and physical actions are not well understood. Current investigations are only adequate to generalize the existence of relationships. Further detailed investigations are necessary to understand the relationships and apply them to the disposal of inactiave tailings and to alteration in the waste technologies of active milling operations.

Considerable insight into the time perspectives of physical instability could be gained by studying playa lake deposits since playa lake beds behave similarly to tailings because of the high salt content. Further detailed studies of selected tailings, based on current experience regarding interactions and locations, could elucidate many of the pertinent questions which have not been asked prior to this investigation. Experimental work could greatly expedite the understanding of processes.

The long-range solution, so as not to generate more problems, seems to require the change of the methods used in the hydrometallurgical processing of the ore. First, a multiple utilization of ore to recover many valuable by-products should be incorporated. Second, salts should be washed out from the waste. Third, the washed tailings should be emplaced dry by separating slime and sand. Fourth, liquid wastes should be evaporated in basins off the ground and not ponded on the tailings surface. And, fifth, sintering or fusing the relatively small amount of slime, if containing radioactive materials, provide a smaller volume which could be disposed of in former mined areas without further barriers analogous to the method described in [13].

Although the problems associated with properties are site specific, physical chemistry provide a common denominator to evaluate processes in tailings materials and interactions with the surrounding environment. More emphasis should be put on approaching evaluations on the basis of physical chemistry instead of the oversimplified approach of looking at "elevated" levels of toxic components or using hydrodynamic methods without a realistic determination of chemical distribution coefficients.

REFERENCES

[1]   Markos, G. and K.J. Bush (1981) Geochemical Investigation of UMTRAP Designated Site at Salt Lake City, Utah. Report UMTRA-DOE/ALO-226-251 [GECR #R-812] (in press).

[2]   Markos, G. and K.J. Bush (1981) Geochemical Investigation of UMTRAP Designated Site at Canonsburg, Pennsylvania. Report UMTRA-DOE/ALO-226 [GECR #R-811].

[3]   Markos, G. and K.J. Bush, Unpublished research notes.

[4]   Bush, K.J. and G. Markos (1981) "Evidence for the Instability of Silicate Minerals in Acid Leach Uranium Mill Tailings". Proceedings of the Fourth Annual Symposium of Uranium Mill Tailings Management. Colorado State University, Fort Collins, Colorado, Oct. 26-27, 1981.

[5]   Havens, R. and K.C. Dean (1967) Chemical Stabilization of the Uranium Tailings at Tuba City, Arizona. U.S. Bureau of Mines Report of Investigations 7288, 12 pp.

[6]   Stumm, W. and J. Morgan (1970) Aquatic Chemistry. An Introduction Emphasizing Chemical Equilibria in Natural Waters. Wiley-Interscience, New York, N.Y.

[7]   Runnells, D. (1969) "Diagenesis Due to Mixing of Natural Waters: A Hypothesis". Nature, v. 224, no. 5217, p. 361.

[8]   Wigley, T.M.L. and L.N. Plummer (1976) "Mixing of Carbonate Waters". Geochim. Cosmochim Acta, v. 40, p. 989.

[9]   Meritt, R.C. (1971) The Extractive Metallurgy of Uranium, Colorado School of Mines Research Institute, Golden, Colorado.

[10]   Markos, G. and K.J. Bush (1981) "Evaluation of Interface Between Tailings and Subtailings Soil, A Case Study: Salt Lake City." Proceedings of the Fourth Annual Symposium of Uranium Mill Tailings Management. Colorado State University, Fort Collins, Colorado. October 26-27, 1981, pp. 135-153.

[11]   Markos, G. and K.J. Bush (1981) "Contamination of Ground and Surface Waters by Uranium Mining and Milling. Vol. II. Field Sampling and Empirical Modeling." Final report to the U.S. Bureau of Mines in preparation.

[12]   Hem, J.D. (1977) "Reactions of Metal Ions at Surface of Hydrous Iron Oxide." Geochemica et Cosmochimica Acta, v. 41, pp 527-538.

[13]   Dreesen, D.R., J.M. Williams, and E.J. Cokal (1981) "Thermal Stabilization of Uranium Mill Tailings". Proceedings of the Fourth Annual Symposium of Uranium Mill Tailings Management. Colorado State University, Fort Collins, Colorado. October 26-27, 1981, pp. 65-79.

THE PRECONCENTRATION OF URANIUM ORES

Henry A. Simonsen
Head : Professional Services Division,
       National Institute for Metallurgy
Randburg, South Africa

ABSTRACT

The dissolution of uranium using acid, or alkaline lixiviants has been almost universally applied in the beneficiation of uranium deposits. The preconcentration of uranium minerals, prior to dissolution, would have important economic benefits in reducing the size of equipment and quantities of reagent required to process low grade ores.

Although preconcentration has been investigated virtually since the inception of large scale uranium beneficiation, the relatively small number of successful installations operated by the mines attests to the difficulties inherent in concentrating minerals that are usually characterised by their small size and disseminative nature. Often the successful implimentation of a preconcentration process depends upon an association between uranium minerals and other minerals with liberation sizes and physical-chemical properties more suited to such processes.

Of the three major ore types comprising some 85 per cent of the Western world's uranium deposits, conglomerates have appeared to receive somewhat more attention as regards preconcentration, particularly where co-product gold is involved. Vein-type deposits, being of somewhat higher grade, are unlikely to provide a discardable tail from any preconcentration routes. Sandstone deposits, due to their good porosity characteristics and the disseminated nature of their uranium mineralization, are more likely to be subjected to such low-cost process alternatives as heap and in-situ leaching.

Processes that have been applied to the preconcentration of uranium ores include:

(a) The use of screens and hydrocyclones to separate uranium minerals on the basis of size/density.

(b) The use of static concentrators such as spirals and Reichert cones.

(c) The use of mechanical separators such as jigs and shaking tables.

(d) The separation of ore and gangue on the basis of density differences using heavy media based separators.

(e) Separation on the basis of magnetic susceptibility using wet high-intensity magnetic separators (WHIMS).

(f) Sorting of feed material on the basis of light-reflectance patterns or radioactivity.

(g) Separation by flotation.

1. INTRODUCTION

Uranium occurrences in the earth's crust range from 1ppm (basalts) to 4ppm (granites). Although the mining of grades of uranium ore as low as 100ppm of U_3O_8 has been suggested economic deposits, without by-product potential, are considered to lie in the approximate range 0,07 to 0,40 per cent U_3O_8 (1). The existence of a co-product such as gold in the case of the Witwatersrand conglomerates can reduce the economic grade to some 0,025 per cent (2). Furthermore, economics of scale can be employed to bring otherwise uneconomic deposits into the workable range. The capital investment required dictates, however, that sufficient ore be available to sustain a profitable period of operation. Such appears to be the case with the Rössing deposit in Namibia where pegmatite ore containing over 100 000 tonnes of U_3O_8 at a grade of about 350ppm is processed in a relatively conventional acid leach plant at a rate of more than 35 000 tonnes per day (3).

Uranium usually occurs as a finely disseminated mineral. This, along with the ready response of most uranium minerals to acid, or alkaline lixiviants, has lead to the almost universal adoption of leach processing of the entire ore. Preconcentration could be expected to make a positive contribution to the process economics of a deposit by reducing the amount of material required to be leached, or by increasing the amount of ore being fed to a mill without a corresponding increase in the downstream capacity of the process.

In practice pre-concentration has had little apparent impact upon uranium beneficiation probably because of the following:

(a) The mineralogical characteristics of most uranium deposits are not conducive towards separation using gravity, magnetic or flotation techniques.

(b) Low cost process routes, such as heap and in-situ leaching, whilst continuing to treat all of the ore, offers an economic advantage in eliminating such large cost centres as comminution and solid/liquid separation.

However preconcentration, a target for considerable research in the early stages of uranium beneficiation process development, continues to receive attention, and in the case, at least, of radiometric sorting, appears to be achieving some success.

2. URANIUM DEPOSITS

The uranium 'ore-type' can be expected to have some influence on its response to processing, particularly as regards the possibility of effective preconcentration. Some 44 per cent of the Western world's uranium occurs in sandstone deposits, 22 per cent in veins, 19 per cent in conglomerates, 8 per cent in shales, the remaining 7 per cent being distributed amongst calcretes, pegmatites, carbonatites, syenites, granites, etc. (4). Some of the characteristics of each of these ore types will be briefly dealt with as follows (5,6):

2.1 Sandstone Deposits

Sandstone ore consists basically of two components, fine to coarse clastic material and a cementing matrix usually containing clay minerals, volcanic ash and organic remains. Uraniferous sandstones often exhibit a high degree of porosity. Grades are generally greater than 0,1 per cent with the uranium occurring mainly in the form of uraninite, pitchblende and coffinite. Where aqueous oxidation has taken place the uranium occurs as hexavalent silicates and oxides, or fixed with vanadate and phosphate to form such minerals as tuyamunite, carnotite and antunite. Carbonaceous material from plant remains is common as are small quantities of sulphides, particularly pyrite. Calcite, montmorillonite clays and chlorite can occur to various degrees in the matrix causing excessive acid consumption during leaching. In the case of the Arlit deposit some 20 to 30 per cent of the uranium occurs bound to a clayey phyllitic phase together with organic matter.

2.1.1 Sandstone Preconcentration Potential

On the face of it the separation of most sandstone deposits into barren clastics and uraniferous matrix components would appear to be possible. In practice the use of physical concentration processes has been limited as reactive gangue constituents tend to report to the fine fraction, where they consume lixiviant, and some uranium accompanies the coarse reject product to waste.

Flotation has been used with some success to separate the feed material into high and low calcite fractions and/or high and low sulphide fractions. In this fashion an optimal combination of acid/alkali lixiviation routes can be employed.

In the case of Arlit ore the uraniferous kaolinite matrix could not be completely separated from the barren silica grains and methods such as attrition scrubbing, gravity concentration, flotation, etc. were unable to produce a discardable tail. Both gravity separation and flotation were tested in the Cosquin uranium deposit in Argentina. Some 70 per cent of the 20 to 25 per cent calcium carbonate content of the feed was rejected. In the process some 75 per cent of the uranium was recovered in a concentrate containing 0,14 per cent U_3O_8 (feed grade ranged from 0,03 to 0,08 per cent U_3O_8).

Radiometric sorting has been applied, with some success, to Karoo sandstone ores.

2.2 Vein-Type Deposits

Vein-type deposits are generally considered to be controlled in their formation by the structural characteristics of the host rock. These deposits characteristically contain significant coarse-grained to massive chunks of uraninite together with more finely dispersed uraninite within the rock matrix. Mineralization is not confined to any particular rock type.

Some deposits have a complex assemblage of metal sulphides (copper, nickel, cobalt, iron) associated with the uranium. In the unoxidized zone uraninite is the dominant uranium mineral with various secondary hexavalent uranium minerals being found above the water table. Because of their variable nature their mineralization and hence preconcentration potential is best dealt with in terms of some individual deposits in Canada and Australia.

2.2.1 Vein Deposit Preconcentration Potential

The Ace-Fay-Verna, Bolger, Nicholson, Martin Lake and Gunnar deposits, in the Beaverlodge area, have been classified as 'classical' vein deposits. Here uranium occurs mainly as pitchblende (with some uranophane in the Gunnar deposit). The Bolgar, Nicholson and Martin Lake ores contain copper, sulphides, cobalt-nickel sulpharsenides, gold, silver and selenides. In most cases the pitchblende is associated with calcite, quartz and minor pyrite. The carbonate content of present feed to the Eldorado-Beaverlodge plant ranges from 5 to 8 per cent Sulphides contained in the Gunnar ore include chalcopyrite, pyrite and galena.

A characteristic of these ores is their hardness (their bond work index is about 20). As the sulphide concentration in these ores would consume an excessive amount of alkali lixiviant, the feed is subjected to flotation. The washed sulphide concentrate is then acid leached with the flotation tailings proceeding to a carbonate-bicarbonate leach.

Those vein deposits described as 'unconformity' types include: Cluff Lake, Rabbit Lake and Key Lake.

The Cluff Lake deposit contains one orebody, the D orebody, that has an average grade of 5 to 10 per cent U_3O_8 with portions assaying up to 65 per cent. Even the lower grade Claude and N orebodies average between 0,5 and 1,0 per cent U_3O_8.

Uranium mineralization includes uraninite, pitchblende and coffinite. Mineralization occurs along shear zones with local concentration near fault intersections. Gold occurs in the D orebody along with cobalt, bismuth, lead and various selenides. Pyrite, galena, pyrrhotite and chalcopyrite occur in all three orebodies.

A limited degree of comminution is found to be adequate (80 per cent -48 mesh). Gravity concentration, when applied to the D orebody, is able to produce a concentrate containing 45 per cent U_3O_8. Apart from possible radiological hazards associated with such a concentration, it is unlikely that a rejectable tailings could be produced.

The Rabbit Lake deposit consists of massive pitchblende infilling fractures and breccia material within a zone of chlorite-quartz-dolomite alteration. This high grade ore is surrounded by a mantle of lower grade material to give an average grade of between 0,3 and 0,4 per cent U_3O_8.

In addition to pitchblende uranium mineralization includes carnotite and uranophane. Gangue minerals include hematite, dolomite, quartz, chlorite and clay minerals. Minor amounts of pyrite, chalcopyrite and galena are present. Although the massive nature of some of the uranium occurrences suggests the application of preconcentration techniques, mine practice has favoured the use of radiometric monitoring as an aid to ore blending.

The uranium mineralization contained in orebodies in the Key Lake area appears to be massive and rich with little gangue mineral in the ore zone. Grades of up to 45 per cent U_3O_8 and 45 per cent nickel are reported for the Gaertner orebody and 20 and 25 per cent respectively for the Deckmann orebody. Uranium occurs as pitchblende and coffinite, nickel as gerodorffite, millerite and niccolite with accessory bravoite. Sulphides including pyrite, sphalerite, chalcopyrite and galena occur in minor amounts.

Again, whereas the massive nature of the mineralization suggests the application of preconcentration techniques, the very high grades involved make the production of a rejectable tail unlikely.

The vein-type deposits in Australia are distributed along a rectangular area, some 100km by 30km, running from north-east to south-west.

The Narbarlek deposit contains almost pure pitchblende in irregular lenses of up to 1 meter in thickness. A more extensive body of lower grade mineralization contains smaller pods of massive mineral and other high-grade material.

The massive high-grade pitchblende is intimately associated with siliceous gangue minerals, particularly mica and chlorite. In the lower-grade primary mineralization, the pitchblende is fine grained and chlorite is a dominant gangue mineral. Sulphides comprise less than 0,5 per cent of the ore.

An oxidized zone, extending from 5 to 15m in depth, contains such secondary minerals as curite, sklodowskite and rutherfordite. Gangue minerals in this zone include kaolinite, quartz, muscovite and chlorite.

The average grade of the orebody is some 2,4 per cent U_3O_8. Raising the cut-off grade from 0,1 per cent to 2,0 per cent would raise the average grade to 9,3 per cent U_3O_8.

The massive nature of the high-grade pitchblende suggests the application of gravity concentration techniques to provide a concentrate containing over 50 per cent U_3O_8. The tailings would be of a high-grade, however, and the processing of same could not be avoided.

The Ranger deposit consists of two orebodies of similar size, though of different depth. Although high-grade intersections are encountered in the orebodies, their general character is somewhat similar to that of the lower-grade primary ore at Nabarlek. At a cut-off grade of 0,05 per cent U_3O_8, the average grade of ore mined will range from 0,22 to 0,25 per cent U_3O_8. Fine grinding is envisaged and preconcentration, with the possible exception of radiometric sorting, is a somewhat less obvious application than in the case of Nabarlek.

The Jabiluka deposit is of somewhat higher grade than that of Ranger with an estimated average of 0,44 per cent U_3O_8 at a cut-off of 0,05 per cent. The two orebodies constituting the Jabiluka deposit together comprise the largest known uranium deposit in the world.

Uranium occurs mainly as uraninite and pitchblende. The host rock series includes graphite schists, chlorite graphite breccia, chlorite schists (with, or without graphite), muscovite chlorite schists, chlorite feldspar schists, dolomite, etc. Although the graphite schist can contain up to 5 per cent pyrite, sulphide grades are lower than those reported for Nabarlek.

Preconcentration, as in the case of Ranger, is unlikely to provide a rejectable tail and would only be considered for cashflow purposes.

2.3 Quartz-Pebble Conglomerates

Uranium-bearing pyritic quartz-pebble conglomerates have been found in Canada, South Africa, Australia and Brazil. Their uranium grade varies from some 200ppm to 0,1 per cent U_3O_8. Their viability is, therefore, considerably enhanced by a co-product potential such as gold in the case of the Witwatersrand deposits.

The conglomerates of the Elliot Lake-Agnew Lake District occur at a shallower depth than those of the Witwatersrand and contain some six times more uranium and are, therefore, viable on the basis of their uranium content alone.

The Canadian conglomerates consist of well-rounded pebbles ranging from 150mm to gravel size which are cemented by a hard crystalline quartzitic matrix. Mineralization occurs mainly within the matrix. The principle uranium mineral is brannerite with lesser uraninite and minor quantities of uranothorite, thucholite, coffinite and gummite. Pyrite and minor amounts of apatite, ilmenite, zircon, rutile and leucoxene occur within the matrix along with quantities of monazite.

The Witwatersrand conglomerate deposits vary from thick beds, similar in appearance to those of Elliot Lake, to thin beds of grit. Uraninite, the principle uranium mineral, occurs in the matrix material as minute crystals (<0,08mm) with varying degrees of roundness. Alteration of individual uraninite grains have occurred due to corrosion during metamorphism and secondary uraninite can be seen enclosing and sometimes replacing primary uraninite grains. Brannerite is a minor uranium mineral and is thought to be the cause of the somewhat lower recoveries of uranium from Witwatersrand conglomerates.

The matrix consists predominantly of quartz with pyrite and other sulphides as well as heavy minerals such as chromite, zircon and leucoxene, the last named being frequently uraniferous due to admixture with brannerite.

Other constituents of the matrix include phyllosilicates, carbon (as thucholite) and minute traces of iridosmine and osmiridium. Uranium associated with thucholite may not respond to conventional acid leach processes.

2.3.1 Conglomerate Preconcentration Potential

As the quartz pebble components of most conglomerates are barren, a separation into pebble and matrix material would appear to be a reasonable starting point for further beneficiation. However, the cohesion of the matrix is such that comminution occurs across the pebbles and liberation of valuable mineral requires the grinding of both pebbles and matrix.

Preconcentration techniques have been extensively tried on both Canadian and Witwatersrand conglomerate ores. In the case of Rio Algom only radiometric sorting, heavy media separation and flotation appeared to offer any potential. The failure of any of these techniques to produce a discardable tailing possibly accounts for the paucity of preconcentration techniques in beneficiation plants in the Elliot Lake area.

In the case of Witwatersrand ores separation by screening, dense media separation, flotation, wet high-intensity magnetic separation, photometric and radiometric sorting have all been tried with varying degrees of success.

2.4 Other Uranium Deposits

One of the more important ore types in this category are the granitic deposits and the most prominent example of the beneficiation of such deposits is that at Rössing in Namibia. In this deposit uraninite is the dominant primary uranium mineral with minor uranium occurrences as betafite. The uraninite is found included in quartz, feldspar and biotite. Beta-uranophane is the most abundant of the secondary uranium minerals.

The occurrence of uranium minerals within cracks in the quartz and feldspar permits a comparatively coarse grind prior to lixiviation (-6 mesh). This same characteristic would inhibit the physical beneficiation of uranium minerals and preconcentration techniques are more likely to be applied to the removal of such acid consuming gangue constituents as calcite rather than to the concentration of the uranium minerals themselves.

Uraniferous pegmatites occur in the Bancroft area of south-western Ontario, Canada. Here the uranium occurs as disseminated, discrete crystals of uraninite and uranothorite occurring along with minor betafite, allanite, zircon and fergusonite within the pegmatite. Uranium grade is generally equal to or less than 0,1 per cent U_3O_8.

The term pegmatite is applied to the Bancroft deposits because of the occurrence of coarse-grained pockets of mineral. Generally, however, their grain size is similar to that of granite. The discrete crystalline nature of the uranium minerals suggests that the possibility of preconcentration cannot be excluded.

Present processing of the carbonatite deposit at Phalaborwa in South Africa does provide an example of the application of preconcentration techniques in the recovery of uranium. Mined primarily for its copper, apatite and vermiculite content the deposit contains small concentrations of uranium and thorium.

The uranium recovery process commences with the desliming of the copper flotation tailings in cyclones, followed by the removal of magnetics and finally by the concentration of a heavy-minerals fraction using Reichert cones. This heavy-mineral concentrate is then processed on shaking tables to separate urano-thorianite and baddeleyite concentrates as well as a tailings product. The uranothorianite concentrate is then leached with nitric acid to remove both the uranium and the thorium.

Uraniferous shales have been considered as deposits of potential economic importance although they are of somewhat low grade. As with other low grade deposits, their eventual exploitation will probably depend upon their co- and by-product potential.

The uraniferous shales of the G. Quantrani Formation in Egypt contain uranium associated with organic material in the presence of dolomite, calcite, gypsum, hematite, goethite, native sulphur, montmorillanite, soluble salts, etc. The grade of the deposit averages some 0,03 per cent U_3O_8 and it has been found possible to effect a degree of uranium concentration using magnetic separation.

The large low-grade uranium deposits, that occur in the bituminous shales in Sweden, have a potential feed grade of between 0,025 and 0,033 per cent U_3O_8. The uranium associated with organic material that occurs in fine grained shale consisting mainly of quartz illite and feldspars. The organic matter comprises some 22 per cent of the shale with a further 13 per cent being made up of pyrite.

In addition to uranium the shale has been considered as a potential source of alumina, potassium, magnesium phosphorous, sulphur, vanadium, molybdenum, nickel and energy. Preconcentration operations have been constrained to the use of heavy-media separation for the removal of limestone prior to further comminution and acid leaching.

The Chattanooga shale formations of the United States have an average grade of 0,004 per cent U_3O_8. Owing to their considerable extent, however, their uranium content has been estimated to be 10×10^6 tonnes of U_3O_8. Again the uranium is associated with organic matter in the presence of aluminium, silicate, iron and sulphur minerals. Preconcentration techniques have been investigated but effective pre-leach treatment appears to be restricted to roasting which is required to increase the porosity of the shale and hence access of the lixiviant.

Calcrete deposits containing uranium have been reported from Australia, Somalia and Namibia. In the case of the Australian deposit at Yeelirrie the uranium occurs as thin coatings of carnotite in cavities and voids comprising a deposit with a grade of 0,15 per cent U_3O_8 in the calcrete. The gangue minerals include quartz, dolomite, calcite, chert, feldspar, gypsum, celestite, clays, etc., an assemblage that precludes considerations of acid leaching.

Preconcentration is unlikely owing to the texture of the gangue minerals. The rejection of sufficient acid consuming minerals to permit the employment of lower-cost acid lixiviant is one possible beneficiation strategy. As grade control is considered to be of some importance, radiometric sorting could have an application.

3. PRECONCENTRATION PROCESSES

Preconcentration (ore-dressing) techniques are likely to be employed in the beneficiation of uranium for one of the following reasons:

(i) To enhance the grade of material prior to leaching with the object of increasing mill throughput without a corresponding increase in installed mill capacity e.g. radiometric sorting of quartz conglomerates.

(ii) To remove minerals that are likely to consume lixiviant e.g. flotation of calcite, or dense media separation of dolomite prior to acid leaching.

Available preconcentration processes include:

(a) Separation on the basis of size and/or shape using screens and/or cyclones.

(b) Separation on the basis of density using dense media, tables, jigs, etc.

(c) Separation on the basis of magnetic susceptibility.

(d) Separation by hand, or machine sorting on the basis of colour or radio-activity.

(e) Separation on the basis of surface properties by means of flotation.

The nature of these various techniques, with some examples of their applications, will be dealt with as follows:

3.1 Separation on the Basis of Size/Shape

The fine dissemination of uranium in most ores suggests that physical separation processes are unlikely to be applied with any efficiency (7). It is possible, however, that uranium would respond more rapidly to comminution than other minerals and therefore lend itself to separation on the basis of size. The tendency of uranium to report in the finer sizes, following comminution, is shown in tables 1 and 2.

TABLE 1: DISTRIBUTION OF URANIUM IN A WITWATERSRAND CONGLOMERATE FEED TO A LEACH PLANT

Size Fraction (microns)	Weight (%)	U_3O_8 (ppm)	U_3O_8 Distribution (%)
+212	0	0	0
-212 +150	0,5	318	0,6
-150 +106	17,7	124	7,3
-106 + 75	9,9	126	4,1
- 75 + 53	11,5	171	6,6
- 53	60,4	404	81,4

TABLE 2: LABORATORY COMMINUTION OF A KAROO SANDSTONE ORE

Size Fraction (microns)	Feed (-10 mesh)			7 min. Grind			12 min. Grind		
	Weight (%)	U_3O_8 (ppm)	U_3O_8 Distrib. (%)	Weight (%)	U_3O_8 (ppm)	U_3O_8 Distrib. (%)	Weight (%)	U_3O_8 (ppm)	U_3O_8 Disbrib. (%)
+600	40,4	676	38,2	8,8	513	6,7	0,6	479	0,4
-600 +300	22,8	722	23,0	32,6	553	26,8	3,3	510	2,3
-300 +150	13,3	664	12,4	18,3	628	17,0	18,3	528	13,4
-150 + 75	8,7	758	9,2	13,3	666	13,2	29,6	592	24,3
- 75 + 38	4,6	792	5,1	8,6	678	8,7	14,4	667	13,3
- 38	10,2	851	12,1	18,4	1009	27,6	33,8	989	46,3

Thus table 1 suggests that a split at 53 microns would place 88 per cent of the uranium in a product comprising only 72 per cent of the feed weight. In the case of Witwatersrand conglomerate deposits, separation by screening is applied, in some cases, after primary crushing to split the feed stream into high and low grade fractions. The high-grade stream is then processed for the extraction of both uranium and gold, the low-grade stream for gold alone. Arrangements can then be made for the separate storage of the low-grade tailings against the event that a more advantageous uranium price would make the reprocessing of same economically viable. Results of some feed-screening processes applied to Witwatersrand conglomerate ores are given in table 3 (8).

TABLE 3: SCREEN SEPARATION OF WITWATERSRAND CONGLOMERATE ORES INTO HIGH-GRADE AND LOW-GRADE FEED STREAMS

Mill	Screen Size (mm)	Undersize	
		Weight (%)	U_3O_8 Distrib. (%)
Blyvooruitzicht	19	30	60
West Driefontein	3	30	60
Western Deep Levels	3	30	60

3.2 Separation on the Basis of Density

Gravity concentration comprises that group of processes in which similar forces produce dissimilar resultants due to the density differences of otherwise similar particles. This group can be subdivided into three sub-groups:

(a) A group in which the separatory force of a flowing film of water is enhanced by the design of the separatory surface. This group includes spirals, plane tables and the Reichert cones.

(b) A second group in which the separatory force of a stream of water is enhanced by mechanical action that promotes the differential settlement/transportation of different mineral particles. This category includes jigs and shaking tables.

(c) A third group in which the density of the fluid medium is so adjusted as to promote the flotation and thus separate removal of the lighter gangue minerals. Into this category fall the dense, or heavy media separators.

Gravity concentration has been extensively employed in the past in the beneficiation of Witwatersrand conglomerate deposits principally to capture coarse gold particles. In 1959, for example, some 43,2 per cent of the gold content of the ore treated was recovered by such gravity concentration devices as corduroy blanket strakes, Johnson concentrators, plane tables, endless belt concentrators and jigs (9).

The fine liberation size of most conglomerate uranium would appear to preclude the concentration of uranium minerals. However, pyrite, gold and other heavy minerals are usually associated with uranium in the conglomerate matrix (10, 11, 12). As the presence of some 8 per cent pyrite in the matrix would increase the specific gravity of the material by 0,1 of a unit, the possibility of gravity separation would be improved if uranium (and gold) were removed as mineral associations with pyrite (13). With the possible exception of relatively massive uranium mineralization, as found in some vein-type deposits, successful concentration by gravity would require the presence of associated heavy minerals.

3.2.1 Static (flowing water) Separators

One of the better known devices in this class is the spiral

concentrator in which the flow of pulp down a spiral subjects particles to centri-
fugal force. The Humphreys, Sala, Reichert and Vickers Xatal spirals have a constant pitch. The Wright spiral has a varying pitch without wash water addition
and the Budin spiral increases in diameter downwards (14, 15, 16, 17, 18).

The application of a spiral concentrator to Witwatersrand conglomerate ores has shown that this device is able to recover some 95 per cent of the
liberated pyrite particles in the size range 53 to 212µm. The overall recovery
of associated uranium is somewhat inadequate, however, as can be seen from table
4 (19).

The Reichert cone separator consists of a number of conical
surfaces provided with annular slots for the removal of concentrate. In testwork
on Witwatersrand conglomerate ore a single cone obtained recoveries of 95, 85 and
60 per cent in the recovery of gold, pyrite (sulphur) and uranium respectively,
albeit at a relatively poor concentration ratio of 2,5 (20).

TABLE 4: TESTS USING HUMPHREYS SPIRAL CONCENTRATION WITH SHAKING TABLE RECONCENTRATION (19)

Feed Material	Grind -75µm (%)	Recoveries (%)					
		In 5% Mass of concentrate			In 10% Mass of concentrate		
		S	Au	U	S	Au	U
Blyvooruitzicht (low-grade)	52,0	61,1	50,2	22,0	63,1	55,0	25,8
Durban Deep Jig Tailing	31,4	76,9	67,5	27,4	80,0	75,4	37,0
" " " "	41,1	69,0	67,0	10,9	77,0	78,6	24,5
" " " "	53,5	63,3	70,2	18,0	69,1	77,6	31,4
" " " "	70,3	59,5	71,2	18,0	63,0	79,4	27,3
Elsburg	50,6	70,8	46,7	21,7	73,3	48,0	27,0
Randfontein Classifier O/F	43,9	57,5	56,2	16,5	59,5	64,4	21,4
Western Areas	49,1	69,1	70,8	27,2	71,9	76,5	31,2
Western Deep Levels	50,7	57,5	65,5	25,4	59,6	69,9	29,1

In the case of the recovery of uranothorianite from the Phalaborwa Igneous Complex tailings from the flotation plant is fed to a Reichert Cone complex comprising:

    6 Rougher Units treating Lo-Phos feed
   18 Rougher Units treating Hi-Phos feed
    2 Scavenger Units treating Lo-Phos tails
    8 Scavenger Units treating Hi-Phos tails
   12 Cleaner Units treating Rougher concs.
    5 Recleaner Units treating Cleaner concs.
    8 Scavenger Units treating Cleaner tails
    1 Quaternary Unit treating Recleaner concs.

The Reichert cone concentrate is further processed using concentrating tables which split the concentrate into three streams namely a uranothorianite concentrate, a Baddeleyite concentrate and a tailings consisting mainly of carbonatite gangue, which is sent back to the scavenger cones. The grades obtained at various stages of the concentration process are shown in table 5(21).

TABLE 5: TYPICAL URANIUM LEVELS IN P.M.C. GRAVITY
           SEPARATION PLANT

Product	U_3O_8 (%)
Rougher Cone Feed	0,0035
Rougher/Scavenger Cone Tails	0,0010
Rougher Cone Conc.	0,0070
Scavenger Cone Tail	0,0013
Cleaner Cone Feed	0,0060
Cleaner Cone Tail	0,0022
Recleaner Cone Feed	0,0170
Recleaner Cone Conc.	0,037
Quaternary Cone Conc.	0,082
Uranothorianite Table Conc.	3,0
Zirconia Table Conc.	0,17

3.2.2 Mechanical Gravity Separators

The two types of separators prominent in this area are the jig and the shaking table. The jig is generally used to treat a feed size range of 25 000 to 75μm whereas the table is applied to the range 3 000 to 15μm (22).

Jigs have been used to recover heavy minerals from Witwatersrand conglomerate ore at such mills as Durban Deep, Western Areas and West Rand Consolidated. Pyrite recovery, however, is generally low (40 to 50 per cent) and the recovery of uranium only one-half of that of pyrite (8).

Gravity methods were applied to high-grade uranium ores in the early days when such ores were processed for their radium content. At the El Sherana mines in the South Alligator area of Australia a small gravity plant using hand-sorting, a jig and a table was used to produce 100 tons of pitchblende concentrate (23).

The shaking table is a fairly efficient concentrator although modifications to deck material and riffle configuration are usually required to optimize performance. As the location of the concentrate band is dependant upon feed composition the use of an automatic product splitter locator would be an advantage. In the case of uranium ores on-line measurement of grade is simplified by the radioactive nature of the ore. Results obtained from the use of an automatic bank splitter in treating Jaduguda uranium ore and uraniferous Rakha copper tailings are compared with results from manual splitter adjustments in table 6 (24).

TABLE 6: SHAKING TABLE BENEFICIATION OF LOW GRADE INDIAN URANIUM ORES

Ore	Control	Product	Weight (%)	Assay U_3O_8 (%)	Distribution (%)
Jaduguda	Auto	Feed	100,0	0,068	100,0
		Conc.	32,0	0,15	70,2
		Tails	68,0	0,03	29,8
	Manual	Feed	100,0	0,068	100,0
		Conc.	24,0	0,17	60,6
		Tails	76,0	0,035	39,4
Rakha	Auto	Feed	100,0	0,016	100,0
		Conc.	8,2	0,15	77,0
		Tails	91,8	0,004	23,0
	Manual	Feed	100,0	0,016	100,0
		Conc.	4,7	0,21	59,5
		Tails	95,3	0,007	40,5

3.2.3 Heavy Media Separators

It has been estimated that a relative density difference of one tenth of a unit is sufficient to permit a heavy medium process to separate ore from waste (25). Sink-float concentration is accomplished by feeding the crushed ore into a fluid body in a vessel provided with devices for the continuous discharge of sink and float particles without excessive loss of the separating medium (26).

The use of a heavy-medium cyclone process has been investigated for the beneficiation of Witwatersrand conglomerate ores. It was found that, when separation was accomplished using a size fraction of between 1,5 and 12,5mm and a relative density of 2,71, some 78 per cent of the uranium could be concentrated into 24 per cent of the mass, a concentration ratio of about 11. Hydrocyclone plants using a ferrosilicon/magnetite mixed media have been installed at Free State Saaiplaas (75 t.p.h.) and Vaal Reefs (80 t.p.h.) with their float products being sent to separate low-grade leach circuits (8).

Heavy medium cyclone concentration tests were conducted on Streitberg Ridge and Armchair Creek ore in Australia. In the case of Streitberg Ridge ore, some 70 per cent of the uranium was recovered in a concentrate comprising some 30 per cent of the feed weight when heating an ore with a head value of 0,055 per cent U_3O_8. Armchair Creek ore, with a head value of 0,1 per cent U_3O_8 provided a recovery of some 65 per cent in some 35 per cent of the feed weight (27).

The application of a heavy medium separatory process to ores of the Uranium City area suggested that between 80 and 90 per cent of the uranium could be recovered in 25 per cent of the feed weight (28). In the case of Elliot Lake ores some 95 per cent of the uranium was recovered in about 70 per cent of the feed weight (29, 30).

3.3 Separation on the Basis of Magnetic Susceptibility

The effective concentration of uranium minerals using magnetic separators has only become a viable proposition with the development of the wet high-intensity magnetic separator (WHIMS). A typical separator of this sort employs an annular box filled with shaped ferromagnetic material and arranged to rotate through one or more high-intensity magnetic fields. Fields in excess of 20 000 gauss are used with a high field gradient being provided by inductive magnetization of the ferromagnetic material (31). A major shortcoming of this form of WHIMS is a tendency for the ferromagnetic matrix to become blocked with wood chips and ferromagnetic particles. Researchers at the National Institute for Metallurgy, in South Africa, have developed a system for the continuous cleaning of the ball matrix of a WHIMS machine.

The recovery of uranium by WHIMS is somewhat dependent upon the particle size of the feed. Thus experiments on Witwatersrand conglomerate residue material showed that although recoveries of up to 84 per cent could be obtained in the size range 106 to 31µm, recovery decreased considerably at sizes below 31µm and were as low as 8 per cent at sizes less than 11µm (32). The results of some WHIMS tests performed on Witwatersrand conglomerate ore, using an Eriez batch separator with a 4,8mm ball matrix, are given in table 7:

TABLE 7: WHIMS BATCH EXPERIMENTS ON WITWATERSRAND CONGLOMERATE ORES

NIM Sample No.	Product	Weight (%)	Assay U_3O_8 (ppm)	Distrib. (%)
G703	Feed	100,0	127	100,0
	Mag.	11,5	771	70,0
	Non-mag.	88,5	43	30,0
G858	Feed	100,0	64	100,0
	Mag.	9,7	419	63,4
	Non-mag.	90,3	26	36,6
G546	Feed	100,0	106	100,0
	Mag.	10,9	660	68,0
	Non-mag.	89,1	38	32,0
G776	Feed	100,0	44	100,0
	Mag.	9,8	329	73,3
	Non-mag.	90,2	13	26,7

Magnetic separation tests conducted on Australian ores produced similar results. Thus the processing of Streitberg Ridge ore (0,055 to 0,065 per cent U_3O_8) produced 60 per cent of the uranium in over 30 per cent of the feed weight, or some 90 per cent of the uranium in over 70 per cent of the feed weight (27).

3.4 Separation by Sorting

The removal of waste rock by sorting is a process of some antiquity and in some instances, as in the processing of Witwatersrand conglomerate ores, the removal of waste rock by hand sorting has accounted for a considerable tonnage of material.

Mechanization of the sorting process permits the application of more sophisticated analytical techniques to large throughput rates. Two types of mechanical sorters have been applied to uranium-bearing ores:

3.4.1 Photometric sorter - where discrimination is based on light-reflectance patterns.

3.4.2 Radiometric sorter - where radioactivity measurement provides a basis for selection.

Photometric sorters have been applied to Witwatersrand conglomerate ores, primarily for gold recovery. At the Doornfontein Gold Mine a Model 13 (ore sorters) was installed to receive a minus 50mm plus 32mm feed (a size range at which manual picking starts to become inefficient). Figures for uranium recovery are not available but the sorter was shown to recover some 80 per cent of the gold in the material fed to it (33). A later development of the Model 13, namely the Model 16 uses a high speed laser scanning technique and is capable of processing some 180t of ore per hour in the size range 8mm to 160mm.

Radiometric sorting was applied to Mary Kathleen ore in Australia in the 1950's. Recent developments have lead to relatively high capacity machines capable of sorting material in the size range 20mm to 160mm. The sorter consists of the following elements:

(a)   A feed presentation system which channels incoming feed material
       into single file streams of spaced rocks.

(b)   A radiation measuring system which uses scintillation counters
       (and optical area measurement techniques) to estimate the grade
       of each rock.

(c)   An electronic processor to compute the U_3O_8 content of a rock
       and to decide whether to accept or reject this on the basis of
       pre-set levels.

(d)   A separation system consisting of a blast manifold that causes
       rocks leaving the belt to fall upon either side of a splitter
       plate.

The response of an ore to radiometric sorting can be assessed on the
basis of radiation measurements and rock area determinations performed on individual rock specimens. Typical results obtained with the aid of a computer
system termed DATAC are given in tables 8 and 9. The response of a Karoo sandstone ore (South Africa) is illustrated by the data in table 10. (43)

TABLE 8: LOW GRADE ORE ZONE - AUSTRALIA (43)
          DATAC results: Estimated feed grade 0,13 kg/t U_3O_8

Accept (%)	Cut/off grade kg/t U_3O_8	Accept grade kg/t U_3O_8	Reject grade kg/t U_3O_8	Recovery U_3O_8 (%)
3,4	0,27	2,38	0,05	62,8
13,6	0,12	0,69	0,04	73,3
21,0	0,10	0,48	0,03	79,5
32,2	0,06	0,34	0,03	86,6
34,8	0,05	0,32	0,02	87,9
36,6	0,05	0,32	0,02	88,6
41,9	0,05	0,28	0,02	90,7
48,2	0,04	0,25	0,02	92,9
52,9	0,04	0,23	0,02	94,3
59,0	0,03	0,21	0,01	95,9
63,3	0,02	0,20	0,01	96,8
68,1	0,02	0,18	0,01	97,7
71,3	0,02	0,18	0,01	98,2
75,3	0,02	0,17	0,01	98,2
79,1	0,01	0,16	0,00	99,2
81,5	0,01	0,16	0,00	99,5
85,0	0,01	0,15	0,00	99,7
89,4	0,01	0,14	0,00	99,9
95,5	0,00	0,13	0,00	100,0
100,0	0,00	0,13	0,00	100,0

TABLE 9: CALCRETE TYPE DEPOSIT (43)
DATAC results: Feed grade 1,24 kg/t U_3O_8

Accept (%)	Cut/off grade kg/t U_3O_8	Accept grade kg/t U_3O_8	Reject grade kg/t U_3O_8	Recovery U_3O_8 (%)
3,6	11,15	11,65	0,85	34,0
8,4	3,4	7,14	0,70	48,1
20,0	2,27	4,39	0,46	70,5
27,9	1,72	3,67	0,31	82,2
31,9	1,36	3,39	0,24	86,7
38,2	0,93	3,02	0,15	92,7
40,2	0,63	2,90	0,13	93,8
46,7	0,47	2,57	0,08	96,6
49,1	0,25	2,46	0,07	97,2
54,4	0,14	2,24	0,06	98,0
60,3	0,13	2,03	0,04	98,6
63,1	0,09	1,95	0,04	98,8
69,5	0,06	1,77	0,03	99,2
72,2	0,05	1,71	0,03	99,3
75,5	0,04	1,64	0,03	99,4
79,3	0,04	1,56	0,03	99,5
82,7	0,04	1,50	0,03	99,6
87,0	0,03	1,43	0,03	99,7
90,0	0,03	1,38	0,02	99,8
100,0	0,02	1,24	0,00	100,0

TABLE 10: RADIOMETRIC SORTING OF KAROO SANDSTONE ORE IN THE SIZE RANGE 1 TO 3 INCHES

Sorter Setting (ppm)	Accept Fraction		
	Weight Cum. (%)	Grade (ppm)	U Dist. Cum. (%)
100 Reject	100,0	601	100,0
300 Reject	42,5	1322	93,5
500 Reject	30,2	1751	88,0
750 Reject	23,0	2132	81,6
750 Accept	15,8	2712	71,1

The radiometric sorter appears to have proved itself in many applications for Witwatersrand conglomerate gold-uranium ores. Two model 17 sorters have been installed at West Rand Consolidated Mines to upgrade -65+25mm material. The tailings from this sorting process, assaying less than 20ppm of U_3O_8, is sent to the waste rock dump.

Nine model 17's are expected to come into operation at Western Deep Levels du ,ng January 1972. Five of these will be 2 channel machines sorting -150+65mm material and a further four will be 5 channel machines sorting -65+25mm material. Reject material is expected, on the basis of pilot plant performance, to assay less than 20ppm U_3O_8 and will be sent to the waste dump.

A 3 channel model 17 has been used in pilot experiments on Vaal Reef ore where its application to -115+65mm material has produced a reject product assaying less than 15ppm of U_3O_8.

At the Welkom Gold Mine, feed material is screened at 6mm to produce high and low grade streams. A -75+50mm fraction of the coarse, low-grade stream has been processed by a 4 channel model 17 radiometric sorter to remove any uraniferous material for processing as part of the high grade stream. A similar application to the -100+63mm fraction of the low-grade stream at Free State Saai-plaas has been experimented with.

At the Buffelsfontein Gold Mine four RM161 6 channel sorters have been applied to -50+25mm run-of-mine ore to provide a reject fraction containing less than 25ppm of U_3O_8.

3.5 Flotation

In the flotation process chemicals are used to selectively render certain mineral particles hydrophobic, thus permitting their attachment to air bubbles and removal from the remainder of the mineral/water mixture. The relative efficiency of this process when dealing with finely ground particles makes it of particular interest for uranium preconcentration due to the finely disseminated nature of many uranium mineral occurrences.

Work on the recovery of uranium from Witwatersrand conglomerate ores was initiated in the 1940's when attention was initially focussed on the use of a fatty acid (oleic acid) scavenging float following the removal of sulphides using xanthate and pine oil as collector and frother respectively (34). At the Government Me-tallurgical Laboratory in South Africa, flotation testwork indicated that (35):

(a) The flotation of thucholite using a frother would recover between 20 and 30 per cent of the uranium.

(b) The addition of xanthate to the float increases the recovery of uranium due to the flotation of uraninite and further thucholite along with the sulphides. Overall recovery amounts to around 54 per cent.

(c) A further increase in recovery can be obtained by adding Aerofloats and sulphonates to the sulphide float. However, the grade declines markedly owing to the flotation of fine gangue minerals.

(d) The maximum recovery of uranium from Witwatersrand conglomerate ore was obtained by adding both oleic acid and xanthate to the float which took place in soft, or moderately soft water. Between 70 and 80 per cent of the uranium was recovered in around 20 per cent of the feed weight.

Early flotation testwork on Elliot Lake conglomerate ore indicated that 92 per cent of the uranium (contained mostly in brannerite, uraninite and monazite) could be recovered in 50 per cent of the weight using 1kg/t of sulphonated sperm oil, 0,5kg/t of Cyanamid 721, 0,13kg/t of crecylic acid and 2kg/t of fuel oil. Continuous pilot plant operation using this reagent combination produced recoveries of 92 per cent in concentrates comprising only 45 per cent of the feed weight, a ratio of concentration of about 2,2:1 (36).

Later testwork, partially aimed at reducing flotation costs by elimina-ting the relatively expensive fatty acids, resulted in a procedure using actinol (a crude, tall oil, fatty acid), in a deslimed, alkaline pulp to recover between 92 and 95 per cent of the uranium in 55 per cent of the feed weight. In pilot plant runs the sulphide was first floated off using 0,08kg/t of xanthate 343 and 0,03kg/t of crecylic acid. The sulphide tailings were then deslimed to provide a rougher uranium flotation concentrate which was subjected to two stages of cleaning using 0,9kg/t Acintol, 0,06kg/t crecylic acid, 0,36kg/t Na_2SiO_3, 0,04kg/t kerosene and 0,3 kg/t Na_2CO_3. In a typical run a flotation feed of 0,15 per cent U_3O_8 would distribute 4,7 per cent of the uranium to the sulphide concentrate, 16,4 per cent to the slimes fraction and 48 per cent to the uranium flotation concentrate. The assay value of the uranium concentrate could be as high as 1,4 per cent U_3O_8.

In a programme to investigate the use of nitric acid dissolution in conjunction with flotation, the Reno Metallurgy Research Centre of the U.S.B.M. employed both conventional sulphide and coal collectors (37). Using sandstone ore and a laboratory flotation cell about 95 per cent of the refractory uranium was recovered with a concentrate to feed ratio of 8:1 from a head grade of 0,25 per cent U_3O_8. Flotation at a pH value of 4,5, required 0,4kg/t of potassium ethyl xanthate and 0,03kg/t of polypropylene glycol methylether.

A uraniferous sandstone at Mindola in Zambia contains uraninite in irregular disseminations with some particles as fine as 15 microns. Flotation experiments were conducted on this ore using soda ash (to give a pH value of 9,4), 0,5 to 0,6kg/t of R633 as a non-carbonate mineral depressant, some 0,6kg/t of palm kernel oil (3 parts mixed with 1 part of Caltex fuel oil) and T.E.B. frother. Later experiments suggested that the addition of 0,25kg/t of sodium silicate would reduce the mica content of the concentrate. It was found that about 80 per cent of the carbonate material in the ore could be rejected with the loss of less than 10 per cent of the U_3O_8 (38).

In an investigation into the flotation of some Australian uranium ores it was found that, although uraninite and pitchblende would respond to flotation using fatty acid collectors, an economically discardable tailings could not be produced (39).

In the case of ore from the South Alligator river area, a fatty acid-hydrocarbon oil emulsion was used to provide the recoveries shown in table 11 below:

TABLE 11: RELATIONSHIP BETWEEN HEAD GRADE AND URANIUM RECOVERY AT AN ENRICHMENT RATIO OF 3:1(CONC. GRADE TO HEAD GRADE)

Head Value U_3O_8 (%)	Recovery (%)
0,25	48
0,53	57
0,78	68
1,25	83

Desliming, prior to flotation, had some beneficial effect. However, the use of gangue depressants did not appear to lead to an overall improvement in flotation. Collector addition amounted to about 0,25kg/t for each 5 minute flotation stage and the pH value was maintained at 6,5.

6. ECONOMIC CONSIDERATIONS

Although both capital and operating costs contribute to the cost of producing uranium oxide, these components cannot be directly related, on a general basis, due to differences in investment allowances, taxation rates, etc.

The approximate distribution of capital and operating costs for various conventional uranium ore processing plants is shown in table 12. There are three ways in which such costs can be reduced:

(i) By eliminating operating steps.

(ii) By modifying equipment, or processes.

(iii) By bringing about an increase in the grade of material processed at as early
       a stage in the process as possible, that is, during preconcentration.

TABLE 12: DISTRIBUTION OF CAPITAL AND OPERATING COSTS IN URANIUM ORE
           PROCESSING (figures rounded off to within 20 per cent of their
           absolute values) (40)

Operation	Capital costs		Operating costs	
	Minimum (%)	Maximum (%)	Minimum (%)	Maximum (%)
Comminution	20	40	15	25
Leaching	5	25	10	25
Solid/liquid separation	15	35	10	20
Concentration and purification	8	20	7	25
Precipitation calcining and packing	4	10	8	15
Tailings disposal	3	10	3	40
Laboratory and services	4	10	6	10

Perhaps the most dramatic examples of the elimination of operating steps are
the use of solution mining (in-situ leaching) and heap leaching. Other processes
in this category include the use of resin and solvent-in-pulp processes.

Examples of technique/equipment modification include the adoption of belt
filters, the use of continuous ion exchange and improvements in the design and
operation of mixer-settlers in solvent extraction.

Unlike most techniques in both of the above categories, preconcentration
usually involves additional capital expenditure and operating costs. Thus any
contribution to improved overall economics must combine the retention of most of
the uranium with a considerable reduction in the mass of material to be further
processed and/or the removal of material (e.g. calcite, sulphides) that would
otherwise contribute to increased processing costs.

Radiometric sorting appears to be making inroads in the area of conglomerate
ore preconcentration and initial pilot scale tests have confirmed its suitability
in the treatment of some sandstone ores. Developments in this field will probably
extend the size range over which radiometric sorting is applicable and hence that
percentage of R.O.M. amenable to sorting.

Wet high-intensity magnetic separation appears to offer some potential in the
upgrading of uranium ores. A comparison of gross profits calculated for the treatment of cyanide residues (from Witwatersrand conglomerate ore) suggests that WHIMS
enjoys an advantage in the lower grade range (see table 13).

TABLE 13: ESTIMATED GROSS PROFIT FROM THE TREATMENT OF CURRENT CYANIDE RESIDUES BY (32):

    (a) Direct acid leaching for U_3O_8

    (b) WHIMS concentration and reverse leaching for U_3O_8 and gold

Grade of U_3O_8 in residue g/t	(1) Gross profit for direct leach of residue c/t	(2) Gross profit for WHIMS followed by reverse leach c/t
165	449	372
150	353	323
125	193	241
100	33	153
95	0	142
76	-	79
52	-	0

Separation on the basis of size is likely to remain of value where the process's economics are likely to benefit from the separate treatment of the coarse and fine products. It is unlikely, however, that size-separation will produce a discardable tail.

Both gravity concentration and flotation will continue to suffer from disadvantageous economics due to their relatively high capital and operating costs and high tailings values. However, as the ERGO operation has shown, a flotation recovery of 35 per cent of the uranium from a head value of 0,006 per cent U_3O_8 can prove profitable when large tonnages are involved (18,6 x 10^6 t.p.a.), co-products are recovered (Au and S) and mining and comminution costs are discounted (treatment of tailings) (42). Heavy media separation has found some application in the separation of conglomerate ore into high and low grade fractions. An early assessment of the economics of heavy media separation on Elliot Lake ore showed that, even with a 90 per cent overall recovery when using HMS plus acid leaching, economics favoured the straight acid leach (see table 14) (41).

TABLE 14: COST OF SINK-FLOAT PLUS ACID LEACHING OF ORE CONTAINING 0,1 PER CENT U_3O_8 (Reproduced from Gow and Ritcey (41)) (Elliot Lake Ore)

Process Step Unit costs are for sink-float plus leaching	Cost Per Day, ($)	
	Sink-Float plus Leaching 3,167tpd mined, 90 per cent over-all recovery	Leaching Only 3,000tpd mined, 95 per cent over-all recovery (Table II)
Crushing @ $0.24/ton	760	750
Sink-float (^{1}) @ $0.30/ton ...	908	-
Grinding (^{2}) @ $0.45/ton	998	1,200
Leaching (^{2})		
Reagents @ $1.00/ton	2,217	3,000
Maintenance @ $0.09/ton	200	210
Labour @ $0.054/ton	120	120
Filtering & thickening (^{2})		
@ $0.35/ton	776	900
Ion exchange)		
Precipitation) (^{3}).......	964	964
Drying & packaging)		
Tailings neut. @ $0.30/ton	950	1,350
Total direct mill costs	7,893	8,494

TABLE 14 (continued)

Process Step Unit costs are for sink-float plus leaching	Cost Per Day, ($)	
	Sink-Float plus Leaching 3,167tpd mined, 90 per cent over-all recovery	Leaching Only 3,000tpd mined, 95 per cent over-all recovery (Table II)
Total direct mill costs ...	7,893	8,494
Indirect mill costs	1,800	1,800
Depreciation	4,140	4,140
Total mill costs	13,833	14,434
Mining, development, etc...	20,585	19,500
TOTAL COSTS	34,418	33,934

COST/lb U_3O_8 RECOVERED

Sink-float plus leaching $= \dfrac{34,418}{5,700} = \$6.04$

Leaching only $ = \dfrac{33,934}{5,700} = \$5.96$

(^{1}) 85% of mill feed = 2,692 tons (35% rejected)
(^{2}) 70% of mill feed = 2,217 tons
(^{3}) These costs proportional to total U_3O_8 recovered.

As far as the impact of preconcentration on the environment is concerned, the following can be said:

(a) In making the processing of residue material economically feasible, preconcentration is likely to reduce the amount of uranium left in tailings dumps.

(b) When applied to a low-grade deposit in order to enhance their profitability potential, preconcentration processes will themselves provide uraniferous tailings material. The amount of uranium contained in such material will depend upon both the inherent efficiency of the process and the manner in which it is operated.

REFERENCES

(1) BOWIE, S.H.U., Global Distribution of Uranium Ores and Potential U.K. Deposits. Geological Aspects of Uranium in the Environment. Geological Society. Misc. paper No. 7, Burlington House, Piccadilly, London, March 1978, p.12.

(2) ibid, p.15

(3) NASH, J.T., Uranium Geology in Resource Evaluation and Exploration. Economic Geology, vol. 73 No. 8, Dec. 1978, p.1404.

(4) JAMES, H.E. and SIMONSEN, H.A., Ore Processing Technology and the Uranium Supply Outlook, Proc. of 3rd Int. Symp. Uranium Institute, London, July 1978, p.157.

(5) BOWIE, S.H.U., The Mode of Occurrence and Distribution of Uranium Deposits. Theoretical and Practical Aspects of Uranium Geology. The Royal Society, London, 1979, pp.35 - 46.

(6)   Significance of Mineralogy in the Development of Flowsheets for Processing
      Uranium Ores. Technical Reports Series No. 196, I.A.E.A., 1980, pp. 109 - 162.

(7)   BURKIN, A.R., The Chemistry of Hydrometallurgical Processes. E. & F.N. Spon
      Limited, 1966, p.7.

(8)   CORRANS, I.J., Comminution and Pre-concentration as Applied to Uranium-Bearing
      Ores. Vacation School: Uranium Ore Processing. National Institute for Metal-
      lurgy, Johannesburg, July 1981, pp. 6.1 - 6.25.

(9)   DOUGLAS, J.K.E. and MOIR, A.T., A Review of South African Gold Recovery Practice.
      Trans. of the Seventh Commonwealth Mining and Metallurgical Congress. Vol. III,
      Johannesburg 1961, p.977.

(10) LIEBENBERG, W.R., The Occurrence and Origin of Gold and Radioactive Minerals
      in the Witwatersrand System, The Dominion Reef, The Ventersdorp Contact Reef
      and the Black Reef. Trans. and Proceedings of the Geological Society of South
      Africa, vol. 58, 1955, pp.101 - 254.

(11) WHITESIDE, H.C.M., Heavy Mineral Analysis of Two Reefs from the Far East Rand.
      M.Sc. Thesis. University of the Witwatersrand, 1943.

(12) LEVIN, J., Concentration Tests on the Gold Uranium Ores of the Witwatersrand
      for the Recovery of Uranium. Jnl. of the S.A. Inst. of Mining and Metallurgy,
      Vol. 57, No. 4, Nov. 1956, pp.209 - 254.

(13) TULT, D.J., The Role of Pyrite in Upgrading the Uranium content of Witwatersrand
      Conglomerate Ores by Means of the Heavy-Medium Separation Process. Jnl. of the
      S.A. Inst. of Mining and Metallurgy, Vol. 70, No. 6, part II. Jan. 1970, pp.195-198.

(14) ANON. The Hymphreys Spiral Concentrator Closed Circuit Test Unit. Manual of
      Operating Instructions Humphreys Investment Co. Bull 10A, Denver, Colorado, 1952.

(15) ANON. The SALA Spiral Concentrator. SALA Maskinfabriks, S. 7 3300, Springfeldt,
      Sweden.

(16) Mineral Deposits Limited. Reichert Spiral Concentrators. Southport, Queensland,
      the Company, Cat. No. RSC1. 1975.

(17) Readings. The Wright Impact Plate Gravity Concentrator. Lismore, N.S.W.,
      Readings, Bull. IPC 475.

(18) ANON. Improved Spiral Concentrators Developed in Austria. Min. J. Lond.,
      Vol. 292, No. 7489, Mar. 1979, p.159.

(19) GUEST, R.N., The Recovery of Pyrite from Witwatersrand Gold Ores. Jnl. of
      South African Inst. of Mining and Metallurgy vol. 76, Oct. 1975, pp.103-105.

(20) GUEST, R.N., A Survey of the Literature on Gravity Separation. N.I.M. Report
      No. 2082, October, 1981, 16pp.

(21) NEL, V.W., VAN DER SPUY, R.C.M. and HESFORD, I.V., Uranium Recovery from
      Uranothorianite Concentrates. Vacation School: Uranium Ore Processing.
      National Institute for Metallurgy, Johannesburg, July 1981, pp.14.ii - 14,39.

(22) TERRIL, I.J. and VILLA, J.B., Elements of High-Capacity Gravity Separation
      C.I.M. Bull vol. 68, No. 757, May 1975, pp.94 - 101.

(23) MURRAY, R.J. and FISHER, W.J., The Treatment of South Alligator Valley
      Uranium Ores, Berkman, et. al. 1968, pp.125 - 137.

(24) NAIR, J.S., DEGALEESAN, S.N. and MAJUMDAR, K.K., Automatic Splitter for Wet
      Tabling of Radioactive Ores. Trans. I.M.M., SectionC, Vol. 83, No. 811,
      June 1974, pp.C121 - 123.

(25) TUTT, D.J., Review of the Application of Beneficiation Processes to Uranium Ores Before Leaching. The Recovery of Uranium. Proc. of a Symposium. São Paulo, August 1970. I.A.E.A.-SM-135/33 pp.33 - 41.

(26) TAGGART, A.F., Elements of Ore Dressing. John Wiley and Sons Inc., 1951, p.169.

(27) GOLDNEY, L.H., CANNING, R.G. and GOODEN, J.E.A., Extraction Investigations with some Australian Uranium Ores. A.A.E.C. Symposium on Uranium Processing, Lucas Heights, July 1972, pp. V.1 - V.18.

(28) GOW, W.A., Concentration of Low-Grade Ace Ores by Heavy Media Separation. Radioactivity Division Special Report SR-136/52, Mines Branch, Department of Energy Mines and Resources, Ottawa, 1952.

(29) HONEYWELL, W.R. and KAIMAN, S., Review of Sink-Float Pre-concentration of Elliot Lake Uranium Ores. Mines Branch Investigation Report IR 69-49, Department of Energy, Mines and Resources, Ottawa, August 1968.

(30) Verbal communication to the April 1962 Meeting of the Canadian Uranium Producers Metallurgical Committee.

(31) BRONKALA, W.J., Magnetic Separation, Mineral Processing Plant Design, Soc. of Mining Engineers (A.I.M.M. and P.E.) 2nd Edit. New York, 1980, pp.467 - 478.

(32) CORRANS, I.J., & LEVIN, J., Wet High-Intensity Magnetic Separation for the Concentration of Witwatersrand Gold-Uranium Ores and Residues. Jnl. of the S.A. Institute of Min. & Met., Vol. 79, March 1979, pp.210 - 228.

(33) KEYS, N.J., GORDON, R.J. and PEVERETT, N.F., Photometric Sorting of Ore on a South African Gold Mine. Jnl. of South African Inst. of Min. & Met., Vol. 75, September 1974, pp.13 - 21.

(34) TAVERNER, L., An Historical Review of the Events and Developments Culminating in the Construction of Plants for the Recovery of Uranium from Gold Ore Residues, Uranium in South Africa 1946 - 1956, Vol. 1 Assoc. Scientific and Tech. Soc. of S.A., Johannesburg, 1957, pp.1 - 19.

(35) LEVIN, J., Concentration Tests on the Gold-Uranium Ores of the Witwatersrand for the Recovery of Uranium. Uranium in South Africa 1946 - 1956, Vol. 1, Assoc. Scientific and Tech Soc. of S.A., Johannesburg, 1957, pp.342 - 387.

(36) HONEYWELL, W.R. and KAIMAN, S. Flotation of Uranium from Elliot Lake Ores, Dept. of Mines & Tech. Surveys, Mines Branch. (Reprint Series RS 3, 1966, 9pp.)

(37) CARNAHAN, T.G. and LEI, K.P.V., Flotation - Nitric Acid Leach Procedure for Increasing Uranium Recovery from a Refractory Ore. R.I. 8331, U.S. Dep. of Interior, Bureau of Mines, 1979, 14pp.

(38) FITZGERALD, M.L. and KELSALL, D.F., Pilot Plant Concentration of Mindola Uranium Ore, Extraction & Refining of the Rare Metals. Paper No. 10. I.M.M., London, 1957 pp.163 - 174.

(39) TRAHAR, W.J., Flotation of Some Australian Ores. Austral. Inst. of Min. & Metal, proc. No. 221, March 1967, pp. 1-9.

(40) SIMONSEN, H.A., BOYDELL, D.W. and JAMES, H.E. The Impact of New Technology on the Economics of Uranium Production from Low-Grade Ores. Fifth Int. Symp. of the Uranium Institute, London, Sept. 1980, 87pp.

(41) GOW, W.A. and RITAY, G.M., The Treatment of Canadian Uranium Ores - A Review. Canadian Min. and Met. (CIM) Bulletin, vol. 62 No. 692, December 1969, pp.1330 - 1339.

(42) WEBSTER, A.S., The ERGO Project: Uranium from Mine Tailings. Sixth Annual Symposium. The Uranium Institute, London, Sept. 1981, 11pp.

(43) SCHAPPER, M.A., The Gamma Sort. - Nuclear Active, Vol. 21, July 1979,
      pp. 11-15.

THE PERMISSION OF THE SOUTH AFRICAN ATOMIC ENERGY BOARD TO PUBLISH THIS PAPER
IS GRATEFULLY ACKNOWLEDGED.

INVESTIGATION OF NITRIC ACID FOR REMOVAL OF NOXIOUS RADIONUCLIDES FROM URANIUM ORE OR MILL TAILINGS

A. D. Ryon, W. D. Bond, F. J. Hurst, F. M. Scheitlin, and F. G. Seeley
Oak Ridge National Laboratory,[*]
Oak Ridge, Tennessee 37830

ABSTRACT

A conceptual process using nitric acid, rather than the currently used sulfuric acid, to extract ^{226}Ra and ^{230}Th in addition to the uranium from ore is proposed in order to decrease the potential hazard from discharge of mill tailings to the environment. Nitric acid leaching of representative uranium ores and tailings from the principal mining districts of the United States removes up to 98% of the ^{226}Ra and ^{230}Th, yielding a residue containing as low as 10 pCi of radium per gram. Leaching of uranium from ores is consistently greater than 99.5%. The residue after multistage leaching with nitric acid is resistant to further radium leaching with water. Radon emanation from nitric-acid-leached residues generally is low due to the low radium content. Heating to 800°C causes further reduction of radon emanation. Greater than 99% recovery of radium from nitric-acid-leach solutions is obtained by carrying on barium sulfate. Good adsorption of radium is also obtained on barite and Celite. Recovery of thorium and uranium by solvent extraction using tri-n-butyl phosphate (TBP) appears promising. Recycle of nitric acid may be accomplished by solvent extraction combined with evaporation and calcination.

---

[*]Operated by Union Carbide Corporation for the U. S. Department of Energy under Contract W-7405-eng-26.

INTRODUCTION

Uranium is usually extracted from ore by leaching with either sulfuric acid or sodium carbonate. These reagents recover >90% of the uranium without extracting significant quantities of radium or other radionuclides in the decay chain of uranium. Disposal of the leached uranium ore residues (tailings) that contain these radionuclides presents a potential problem. Millions of tons of tailings now exist at 24 abandoned sites where uranium mills operated during the period 1948-1970. In addition, fifteen mills are currently in operation with a total daily capacity of 25,000 metric tons of ore. The radionuclides posing the major potential health hazards are ^{226}Ra and its daughter ^{222}Rn. Pathways may occur for the solubilization of ^{226}Ra in water supplies (MPC = 5pCi/L) [1] and/or the contamination of air with gaseous ^{222}Rn. Other radionuclides of importance are ^{230}Th (the parent of ^{226}Ra), ^{210}Pb, and ^{210}Po. The latter two are relatively short-lived and would soon decay if their long-lived parent ^{226}Ra were removed.

An engineering analysis of methods for treating milling wastes, including tailings from a uranium mill, was made in 1975 at ORNL as a part of the "as low as reasonably achievable" (ALARA) disposal study [2]. The study evaluated various treatment methods for tailings to determine both the costs and the benefits from the reduced radiation dosage to man and environment. One of the advanced and more speculative methods considered was the substitution of nitric acid for sulfuric acid to leach out most of the radionuclides of interest, such as radium, thorium, lead, and polonium, in addition to the uranium. The process was based on encouraging data obtained in a few scouting tests [3]. Subsequently, other investigators made tests using hydrochloric acid [4-5].

Based on the scouting tests, a conceptual flowsheet (Figure 1) was devised for nitric acid leaching of uranium ore to dissolve uranium, radium, and thorium. In this process, the leached solids are washed extensively by countercurrent decantation (CCD) to prevent discharge of nitrate and soluble radionuclides. The pregnant liquor is then treated to remove radium (usually by barium sulfate carrier), the uranium and thorium are removed by solvent extraction, and the nitric acid is recycled by evaporation and calcination of the remaining nitrate salts. Results of recent laboratory tests [6,7] are presented as a basis for the flowsheet. Considerable emphasis has been placed on leaching of radium, in addition to the scouting tests for leaching of thorium, uranium, polonium, and lead. Some characterization of the nitric-acid-leached tailings has been done in order to determine radon emanation and leachability of the residual radium. Several alternative methods for the recovery of radium, uranium, and thorium from the leach liquors have been explored.

ORNL DWG 81-15611

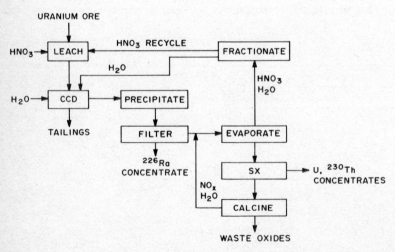

Figure 1 Conceptual Flowsheet for Nitric Acid Processing of Uranium Ore

LEACHING TESTS

Samples of ore feed and the corresponding tailings from five operating mills in the western United States were used in most of the tests. Scouting tests for leaching of radium with nitric acid showed several results that were generally applicable to both ores and tailings. Equilibrium was attained within 0.5 h, and further contact (up to 48 h) does not increase leaching. Single-stage leaching of radium increased from 30% to 70% as the concentration of nitric acid was increased from 1 to 8 molar. This is consistent with the solubility behavior of other alkaline earth sulfates, particularly barium sulfate [6]. Higher temperatures gave increased leaching; consequently most of the data were measured at temperatures ranging from 60 to 80°C. Although crosscurrent, multiple-stage leaching was required to obtain effective extraction of radium at the conventional solids content of 33%, equivalent leaching was obtained using 14% solids in a single stage. The results for one ore sample using 3 $\underline{M}$ HNO$_3$ (Figure 2) show that most of the radium is leached in two stages at 30% solids. The fraction removed in subsequent

ORNL—DWG 79—1025R

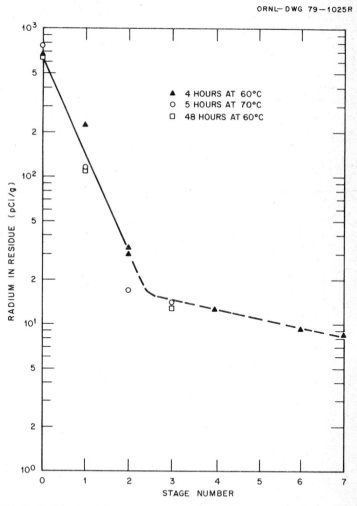

Figure 2 Radium Leaching from Uranium Ore with 3 $\underline{M}$ HNO$_3$ at 30% Solids

stages is significantly lower and is nearly constant, indicating that the bulk of the radium probably exists as a sulfate or carbonate and that the remaining portion probably is in a refractory mineral or is strongly adsorbed. Since additional residence time of 48 h (for one test) did not increase leaching, the leaching apparently is not limited by diffusion. The radium content was reduced to <10 pCi/g with seven leaching stages.

Significant radionuclides leached by two stages with nitric acid from representative ore samples are summarized in Table I. Leaching of ^{226}Ra was >93% for all samples except ore #5, a high lime ore from an alkaline leach mill, which gave 85% leach results. Leaching of ^{230}Th was consistently higher (96-99%). Leaching of ^{210}Po and ^{210}Pb was variable and generally lower than for radium. Leaching of uranium was consistently excellent (99.5% or greater), offering an important advantage over the lower leaching (91%) by the conventional sulfuric acid process. This may be especially important for the lower-grade ores that will be milled in the future.

TABLE I. Two-stage, Crosscurrent Leaching of Significant Radionuclides from Ores with 3 M HNO$_3$ at 33% Solids for 5 h at 70°C,

Ore No.	^{226}Ra (%)	^{230}Th (%)	^{210}Po (%)	^{210}Pb (%)	U (%)
2	94	99	82	62	99.7
3	93	96	67	61	99.7
4	98	98	86	73	99.7
5	85	96	85	25	99.6
6	97	97	49	56	99.5

CHARACTER OF NITRIC-ACID-LEACHED RESIDUES

As shown in Figure 2, the radium content of the residue can be reduced to less than 10 pCi/g by multistage leaching. Although this does not meet the proposed EPA standard (5 pCi of ^{226}Ra/g) for contaminated soil, it does meet a United Kingdom standard [4] of 25 pCi/g for soil and building materials. Since the EPA standard is only slightly higher than the natural radium concentration in soil and is currently being considered for possible revision, it may be possible that the nitric-acid-leached residue would meet radium throw-away standards and would not require costly disposal methods.

The mass distribution in the sand fraction (+140 mesh) of the residues from two-stage nitric acid leaching of ore and tailings ranged from 33 to 88% (Table II). Even though the concentration of radium in the slime fraction of the residues was about twice that in the sands, the fraction of total radium was approximately equal in each fraction. This differs markedly from sulfuric acid tailings where most of the radium occurs in the slime fraction, probably associated with calcium sulfate or adsorbed on the small particles.

The water leachability of radium from mill tailings is an important problem in disposal, particularly for a wet climate or if disposal is to be in or near an aquifer. The radium concentration leached by distilled water is nearly 3 orders of magnitude lower for a nitric-acid-leached residue than for a sulfuric acid tailing (Table III). Furthermore the concentration of radium is only 3 pCi/L, which is lower than the standard for drinking water.

The nitric acid process has obvious potential for nitrate contamination of the environment, if the tailings contain nitrate. Our standard leach test procedure included filtering to separate the residue from the acid leach solution, repulping the residue with water, and refiltering. A 3% slurry of the washed residue in distilled water showed less than 2ppm nitrate, which is less than the drinking water standard. It is believed that a mill with a good wash circuit could produce tailings that would meet the standards for both soluble radium and nitrate. An alternative to thorough washing would be the use of bacterial treatment to destroy the nitrate; this has been demonstrated on a pilot scale [8].

Probably the most important benefit of the nitric acid process is the low radon production resulting from the low concentration of the ^{226}Ra parent in the residue. The radon emanation coefficient (Table IV) varies over a wide range for ores, mill tailings, and nitric acid residues. Surprisingly, the emanation coefficient for the nitric acid residues is not lower than that for ores or mill tailings, even though only the refractory radium is left. The reduction in radon flux from the residual nitric acid process tailings is therefore directly dependent on the reduction of radium content in the residue. Removal of 95% of the radium causes the radon flux to be decreased proportionately.

TABLE II. Radium Distribution in Sand and Slime Fractions of Residue from Two-Stage 3 $\underline{M}$ HNO3 Leach at 70°C with 33% Solids for 5 h

Source	Sand fraction[a] (wt %)	^{226}Ra Concentration (pCi/g)		$\frac{Slime}{Sand}$	Fraction of Total Ra In sand (%)
		Sand[a]	Slime[b]		
No. 1 tailings	53	42	59	1.4	45
No. 2 ore	82	31	58	1.9	71
No. 3 ore	66	39	92	2.4	45
No. 4 ore	74	13	31	2.4	57
No. 5 ore	88	35	109	3.3	70
No. 6 ore	33	53	39	0.7	40

[a]+140 mesh (>105 μm).

[b]-140 mesh (<105 μm).

TABLE III. Radium Leaching with Distilled Water (3% solids for 6 days)

Source	Radium Concentration	
	Solid (pCi/g)	Water (pCi/L)
H2SO4 tailing	654	1620
HNO3 residue	9	3

Further reduction of the radon emanation was obtained by heating the residues; at 800°C the emanation coefficients were reduced to about one-third of the original values. Similar reductions were also obtained with sulfuric acid tailings. It is thought that annealing of crystals at this temperature [9] repairs the damage caused by alpha decay of radionuclides, which probably explains the reduction in radon emanation.

TABLE IV. Radon Emanation Coefficient

Source	As Received		Nitric Acid Residue	
	Ore (%)	Tailings (%)	Ore (%)	Tailings (%)
1	-	10	-	22
2	28	17	8	8
3	26	12	15	14
4	20	9	43	30
5	35	45	17	21
6	9	14	38	37

RECOVERY OF RADIONUCLIDES

The recovery and isolation of ^{226}Ra, ^{230}Th, uranium, and possibly ^{210}Po and ^{210}Pb from the leach liquor is an important part of the proposed nitric acid process. Several scouting tests have been made to determine the feasibility of recovery of radium, thorium, uranium, and nitric acid. Much more research is needed to optimize the recovery steps to attain a degree of isolation or concentration of the recovered radionuclides sufficient to permit efficient storage.

A test was made of radium recovery by the classical method of carrying on barium sulfate. The effects of various concentrations of barium and nitric acid on the radium recovery are shown in Figure 3. Good recovery (>99%) is obtained with a barium concentration of 10 mM for nitric acid concentrations up to 1 M. Radium recovery decreases at lower barium and/or higher nitric acid concentrations in the presence of excess sulfate. This quantity of barium would yield a Ba-Ra product equivalent to approximately 9 kg per metric ton of ore processed. Further concentration of this product would be needed for transportation and storage away from the mill site.

Several alternative methods of radium recovery were also tested. One method used oxalate precipitation of thorium, calcium, and lead. The thorium oxalate was very insoluble even in 3.6 M nitric acid, but it did not carry radium in nitric acid at concentrations greater than 0.2 M. Calcium and lead oxalates were too soluble in dilute nitric acid to be useful. The adsorption of radium obtained on barite from 0.5 M and 3 M nitric acid and on Celite (diatomaceous earth) from 0.5 M nitric acid was encouraging.

The recovery of ^{230}Th and uranium was examined using solvent extraction. Table V shows the results of batch equilibrations of leach solution with an organic solution of tri-n-butyl phosphate (TBP), which is commonly used for reprocessing nuclear reactor fuel. Good recovery of uranium and thorium was obtained in four stages with a reasonable ratio of solvent to leach liquor. Calculations made with a computer code (SEPHIS) [10] indicate that fractionation by solvent extraction will produce (1) a product containing >98% of the uranium (only slightly contaminated with thorium); (2) a product containing 80% of the thorium (and about 1.5% of the uranium); and (3) a product containing 75% of the nitric acid and 20% of the thorium. It is believed that this nitric acid can be recycled to the leach circuit. The remainder of the nitric acid and traces of uranium and thorium would be in the raffinate stream.

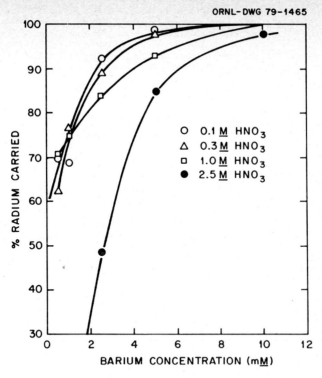

ORNL-DWG 79-1465

Figure 3 Carrying of Radium on BaSO4 as a Function of Barium Concentration in Various Concentrations of Nitric Acid

TABLE V. Batch, Crosscurrent Extraction of Uranium and ^{230}Th

Leach Liquor[a] Composition: 2.5 $\underline{M}$ HNO$_3$, 1430 mg U/L and 572 pCi ^{230}Th/mL
Solvent Composition: 30% TBP—70% $\underline{n}$-Dodecane

Stage No	Phase Ratio, Organic/ Aqueous	Uranium		D[b]	Thorium		D[b]
		Organic (mg/L)	Aqueous (mg/L)		Organic (pCi/mL)	Aqueous (pCi/mL)	
1	1.0	1360	50	27	158	198	0.80
2	1.0	40	2	20	72	117	0.62
3	1.0	1.9	0.09	21	30	79	0.38
4	5.0	0.025	<0.02	-	8.1	13.1	-

[a]Obtained from a 3 $\underline{M}$ HNO$_3$ leach of No. 4 uranium ore.

[b]D = distribution coefficient = ratio of the concentration in the organic phase to the concentration in the aqueous phase.

It is proposed that this stream be evaporated and the nitrates calcined to oxides for disposal (see Figure 1). The oxides of nitrogen would then be scrubbed and recovered for recycle to the leach circuit. A single calcination test at 600°C showed that the nitrate content of the calcined solids was 0.5%. Adsorption of nitrogen oxides from air has been studied and shown to be feasible [11]; this may have the potential of producing nitric acid for recycle to leaching.

ESTIMATED COST

The cost estimate for producing U_3O_8 by the conceptual nitric acid process that was published in the ALARA report [12] was updated to provide guidance for further work. A cost comparison with a conventional sulfuric acid mill was made by estimating the additional capital and operating requirements for the nitric acid process. It must be recognized that the cost estimates shown in Table VI are highly uncertain because of the lack of data for the conceptual process. Most of the estimated additional capital cost was for equipment such as evaporators, fractionators, and calciners that are required to recycle the nitric acid. Also all equipment must be stainless steel, rather than wood or plastic, in order to contain the corrosive acid. More than 50% of the estimated annual operating cost was for the fuel required for evaporators and calciners. The additional cost for the nitric acid process was estimated to be $16.70 per kg of U_3O_8. If allowance is made for increased uranium recovery with nitric acid and a price of $66 per kg of U_3O_8, the incremental cost is reduced to $9.90 per kg of U_3O_8.

TABLE VI. Estimated Additional Costs for the Conceptual Nitric Acid Process[a]

Capital costs	$36 × 10^6
Annual cost (Including fixed)	$20 × 10^6
Cost per kg of U_3O_8	$16.90
Adjusted cost[b]	$9.90

[a]Cost estimates are for additional capital and operating requirements for the nitric acid process, compared to the cost for a conventional sulfuric acid process at a mill processing 1818 metric tons of ore per day (1980 dollars).

[b]Allowing for an increase in uranium recovery from 91 to 99% and a price of $66 per kg of U_3O_8.

CONCLUSIONS

The use of nitric acid rather than sulfuric acid in uranium mills offers the possible benefits of (1) reduction in the potential radioactive hazards of the solid tailings and (2) increased uranium extraction. More than 90% of the ^{226}Ra and ^{230}Th can be leached from the ore with the nitric acid process. The remaining ^{226}Ra would produce only ~10% as much ^{222}Rn as conventional tailings and would therefore reduce the amount of covering required to meet radon regulations. The residual radium is virtually insoluble in water and would probably not result in contamination exceeding drinking water standards. Dissolution of uranium is >99.5%, which is particularly significant for the lower grade ores that will be milled in the future. The proposed solvent extraction circuit utilizes TBP, which is used in uranium refining, thus offering the potential (with additional purification cycles) for producing refinery-grade uranium at the mill.

Several problems are associated with the nitric acid process. Most importantly, the cost is greater than that for conventional processes. Adequate storage of the radium and thorium concentrates must be provided although, if sufficient concentration can be accomplished, they may be stored with other alpha-emitting wastes such as the transuranics. Nitrate from the use of this process is also a possible contaminant that must be contained in the mill and thoroughly removed from solid tailings. Although the conceptual flowsheet appears to be technically feasible, more research is needed to develop the process and provide a sound basis for a reliable cost-benefit analysis.

REFERENCES

1. Interim Drinking Water Standards, Title 40, CFR Chap. 1, Part 141 (published Dec. 24, 1975; scheduled to be effective June 24, 1977).

2. Sears, M. B. et al., Correlation of Radioactive Waste Treatment Costs and the Environmental Impact of Waste Effluents in the Nuclear Fuel Cycle for Use in Establishing "As Low As Practicable" Guides — Milling of Uranium Ores, ORNL/TM-4903, Vol. 1 (May 1975).

3. Seeley, F. G., "Problems in the Separation of Radium from Uranium Ore Tailings," Hydrometallurgy 2, 249-63 (1977).

4. Borrowman, S. R., and Brooks, P. T., Radium Removal from Uranium Ores and Mill Tailings, Bureau of Mines Report of Investigations 8099, Salt Lake City Metallurgy Research Center (1975).

5. Saint-Martin, N. and Haque, K. E., "Hydrochloric Acid Leaching from Elliot Lake Uranium Ore — A Preliminary Study", Division Report ERP/MSL 77-339 (TR); CANMET, Energy, Mines and Resources Canada; 1977.

6. Ryon, A. D., Hurst, F. J., and Seeley, F. G., Nitric Acid Leaching of Radium and Other Significant Radionuclides from Uranium Ores and Tailings, ORNL/TM-5944 (August 1977).

7. Scheitlin, F. M., and Bond, W. D., Removal of Hazardous Radionuclides from Uranium Ore and/or Mill Tailings: Progress Report for the Period Oct. 1, 1978 to Sept. 30, 1979, ORNL/TM-7065, 1980.

8. Pitt, W. W., Hancher, C. W., and Patton, B. D., "Biological Reduction of Nitrates in Wastewaters from Nuclear Processing Using a Fluidized-Bed Bioreactor", Nuclear and Chemical Waste Management, Vol. 2, Pergamon Press Ltd., London, (1981).

9. Faul, H., Nuclear Geology, Wiley, New York, 1954.

10. Mitchell, A. D., Modification of the Sephis — Mod. 4 Computer Program to Simulate the Thorex Solvent Extraction Process, ORNL/TM-6825, (1979).

11. Counce, R. M. and Groenier, W. S., Nitrogen Oxide Absorption into Water and Dilute Nitric Acid — Descriptions of a Mathematical Model and Preliminary Scrubbing Results Using an Engineering-Scale Sieve-Plate Column, ORNL/TM-5910 (June 1978).

12. Ryon, A. D. and Blanco, R. E., Correlation of Radioactive Waste Treatment Costs and the Environmental Impact of Waste Effluents in the Nuclear Fuel Cycle for Use in Establishing "As Low as Practice" Guides— —Appendix A. Preparation of Cost Estimates for Volume 1, Milling of Uranium Ores, ORNL-TM-4903, Vol. 2, p. 119, (May 1975).

CHLORIDE METALLURGY FOR URANIUM RECOVERY:
CONCEPT AND COSTS

by M.C. Campbell, G.M. Ritcey and E.G. Joe
CANMET, Energy, Mines and Resources Canada
Ottawa, Ontario

ABSTRACT

Uranium, thorium and radium are all effectively solubilized in chloride media. This provides a means to separate and isolate these species for ultimate sale or disposal. The laboratory work on the applications of hydrochloric acid leaching, chlorine assisted leaching and high temperature chlorination is reviewed. An indication of costs and benefits is provided to enable the evaluation of this technology as an option for reducing the environmental impact of tailings.

BACKGROUND

The Canada Centre for Mineral and Energy Technology (CANMET) is the principal federal government metallurgical laboratory in Canada. The close association with the uranium industry goes back over three decades when the Mines Branch developed the uranium extraction technology applied in Canada's fledgling uranium industry. Uranium is somewhat of a special commodity in CANMET because of this long association and also because the federal government has a special interest in this controlled commodity.

The Canadian uranium industry has had its triumphs and disappointments, but in 1981 it is a reasonably healthy industry working generally profitably in seven primary plants plus one phosphoric acid plant.

Canadian uranium mills have produced in excess of 100 million tonnes of tailings to date and are adding to these tailings at a rate of 20,000 tonnes a day. Doubts have been expressed about the management of these tailings particularly after mill shutdown. In Canada we are by no means certain of the scope of the long term management problems and have therefore initiated a National Tailings Research Program, a description of which will soon be publicly available.

Conventional sulphuric acid (and alkaline) leaching of uranium ores has been the basis for the successful and efficient extraction of uranium in plants throughout the world. Uranium extractions exceed 90% and often 95%. However these processes dissolve less than 5% of the radium-226 so that the bulk of this radionuclide is discharged in tailings solids[1]. Subsequent weathering and the leaching action of water results in solubilization of radium to levels exceeding environmental regulations. Current practice is to add barium chloride to the effluent flowing from the primary tailings pond in order to precipitate the radium as a mixed barium radium sulphate. Although this practice provides for safe disposal and the discharge of final effluents that meet environmental regulations, treatment after mine closure will be expensive to monitor and control. Furthermore, most Canadian uranium ores contain sulphides which are not altered by the uranium extraction process. These sulphides are discharged in the tailings combined with spent sulphuric acid liquor. These tailings are neutralized with lime to pH 10.5 prior to discharge to large tailings impoundments. Over time (as little as several years), the lime is weathered and an acid drainage problem arises in the tailings dam seepage and runoff.

INTRODUCTION

Because of the apparent inadequacies of present technology to avoid long term environmental impact, a research program was undertaken at CANMET on the development of alternative technology.

The research objectives for the project are stated as follows[2].

"To effect minimum environmental impact and resource conservation by the development of alternate process technology to the use of sulphuric acid, such that the recovery of uranium and by-products of thorium and rare earths, nickel and arsenic, will be increased, radionuclides will be solubilized for subsequent isolation and disposal, and minimum sulphides (and arsenic) will be disposed in the tailings. Any process has to be economically acceptable."

Very early in the project planning chlorides and nitrates were considered as the possible leachants to solubilize the radionuclides for subsequent isolation (e.g., Ra^{226}, Th^{230}, Pb^{210}). A target of less than 25 pCi Ra^{226}/g in the tailings residue was set as the objective. Such a strong leachant would have the advantage that all sulphides would be decomposed during the process, with no subsequent acid drainage problem in the tailings. Furthermore acid could be recovered for recycle, thus decreasing the large tailings storage areas and reducing the neutralizing costs. From such leach liquors it would also be possible to recover valuable by-products of rare earths and thorium from the Elliot Lake ores, and nickel and arsenic from the Saskatchewan ores.

The project on new technology for uranium extraction began with a critical assessment of the literature by CANMET staff[3] with an objective of determining any existing technology for uranium ore treatment whereby the radium-226 would be quantitatively solubilized in the leach process. Conventional H_2SO_4 leaching was known to be inadequate[1]. A nitric acid route was reported to be somewhat more successful[4,5] but it was not effective with all types of ore. Chloride leaching of uranium ores had been developed but no data were reported on radium-226 solubility. Chloride technology appeared to present an obvious challenge and a possible solution.

The literature survey[3] indicated two possible chloride routes, one an anhydrous chlorination process and the other an aqueous hydrochloric acid leach process. Subsequently a literature survey was conducted on the leachability of various radioactive minerals in different leachants[6]. Both aqueous and anhydrous chloride processes were selected for the subsequent con-current investigations.

RESEARCH WORK

The initial work was on the typical low grade Canadian uranium ores of Elliot Lake. Subsequent research was conducted on the high grade complex arsenical ores of Saskatchewan.

The Elliot Lake ore consisted of a coarse-grained quartz-pebble conglomerate[7]. The ore analysis is presented in Table 1.

The uranium-bearing minerals are brannerite, which is a uranium titanate [(UO, TiO, UO_2) TiO_3], and uraninite [UO_2]; the thorium is believed to be associated with monazite, which is a rare-earth phosphate [(Ce,La)PO_4][7].

The activity values for Ra-226 and Pb-210 are close to those expected for a uranium ore in secular equilibrium with its daughters, since each daughter should have an activity of 331 pCi/g for every 0.10% U in the ore.

Table 1 Some constituents of Canadian uranium ores

| Elements | Concentration (Wt %) | | | | |
| | Ontario | | | Saskatchewan | |
	Elliot Lake	Agnew Lake	Madawaska	Rabbit Lake	Key Lake
U	0.14	0.031	0.09	0.40	1.50
Th	0.036	0.24	0.036	0.002	0.002
Ra^{226}	400 pCi/g	225 pci/g	265 pCi/g	1250 pCi/g	5220 pCi/g
Fe	3.12	2.52	3.46	3.09	3.04
S	2.87	1.0	1.0	0.40	0.78
Si	38.39	37.3	27.0	25.0	26.0
Al	3.62	4.32	6.14	7.64	7.68
Mg	0.12	0.34	1.48	8.31	2.17
Ca	0.14	0.30	4.90	1.40	0.37
P	0.40	0.540	0.040	0.08	--
Ti	0.22	0.18	--	--	0.54
C	--	--	--	0.9	0.36
CO_2	--	--	--	2.29	0.57
Cu	--	--	--	--	0.04
As	--	--	0.001	0.02	1.23
Co	--	--	--	0.01	0.05
Ni	--	--	0.02	2.06	
RE	0.15	--	--	--	--

Chloride Leaching

Initially some preliminary bench scale aqueous HCl leach tests were performed, resulting in the dissolution of 91% U, 85% Ra[226] and 63% Th[8]. Those results were sufficiently encouraging to continue research on that route. Both HCl[9] and Cl[10] assisted aqueous leaching processes were developed. The results of the HCl leaching are summarized in Table 2[9]. The ore grind was 50% minus 200 mesh, and leaching was at 50-58% solids at a temperature of 75°C for 18 hours. Two-stage leaching was effective in reducing the radium-226 content to a level of 10-25 pCi/g, by leaching in the second stage for 6 hours with HCl or Cl$_2$ additions.

Table 2 Two stage chloride leach results (Elliot Lake ore)

Leach type	Stage No.	Overall extraction (%)			Final residue analysis		
		U	Th	Ra	U(%)	Th(%)	Ra[226](pCi/g)
HCl	1				0.04	0.026	196
HCl*	2	98	79	95	0.0027	0.0076	20
HCl	1				0.026	0.03	105
Cl$_2$**	2	99	83	95.4	0.002	0.006	18

* 6 h retention time, 75°C, 2.0 kg NaClO$_3$/tonne

**6 h retention time, 80°C, 12.7 kg Cl$_2$/tonne

Chlorine-assisted leaching was also applied to several ores[10]. The conditions and results of optimum two-stage leaching to date are shown in Tables 3 and 4. However, the complete solubilization of Ra[226] on the high grade complex ores has not yet been achieved, and requires further investigation.

Table 3 Optimum leach* conditions for Cl$_2$HCl 2-stage** leach
of uranium ores

Stage	Feeds				
	Elliot Lake	Agnew Lake	Madawaska	Key Lake	Rabbit Lake
First stage Temp (°C)	80	80	60	90	90
Time (h)	18	10	10	10	10
Cl$_2$(kg/tonne)	16.0	20.0	16.0	23.0	28.0
Final mV	450	800	1000	550	700
Final pH	1.0	1.5	1.5	1.5	3.5-4.5

* Slurry density and the particle size for all ores were 50% solids and 60% minus 200 mesh.

**Second stage leach of the first stage residue conducted with 44 kg HCl/tonne at 65°C for 6 hours.

The results thus indicated that HCl or Cl$_2$ were effective for the dissolution of U and Th as well as being effective for Ra[226] solubility, resulting in tailings containing F=L pCi Ra[226]/g for most ores. A conceptual flowsheet for HCl leaching is shown in Figure 1. Note that the flowsheet shows two stages of

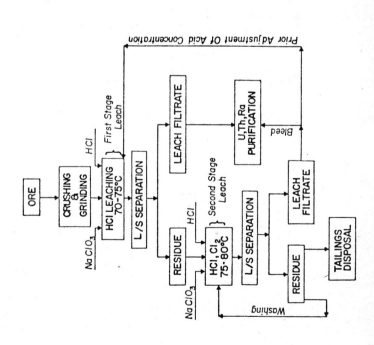

FIGURE 2

SHAFT FURNACE CHLORINATION

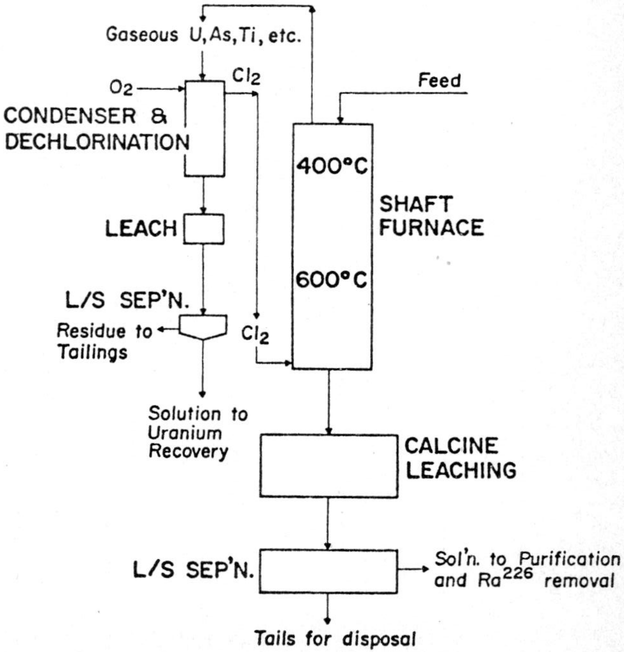

Feed

SHAFT FURNACE

400°C

600°C

Gaseous U,As,Ti, etc.

Cl_2

O_2

CONDENSER & DECHLORINATION

Cl_2

LEACH

L/S SEP'N.

Residue to Tailings

Solution to Uranium Recovery

CALCINE LEACHING

Sol'n. to Purification and Ra^{226} removal

L/S SEP'N.

Tails for disposal

FIGURE 1

TWO-STAGE LEACH WITH HYDROCHLORIC ACID

Prior Adjustment Of Acid Concentration

ORE

CRUSHING & GRINDING

HCl

$NaClO_3$

First Stage Leach

HCl LEACHING 70-75°C

L/S SEPARATION

LEACH FILTRATE

U,Th,Ra PURIFICATION

Bleed

RESIDUE

HCl

Second Stage Leach

HCl, Cl_2 75-80°C

$NaClO_3$

L/S SEPARATION

LEACH FILTRATE

RESIDUE

TAILINGS DISPOSAL

Washing

Table 4   Solubility after Cl_2-HCl leaching

| Ore feed | Stage | Cumulative extraction (%) | | | | | Residue |
		U	Th	Ra^{226}	Ni	As	(pCi/g Ra^{226})
Elliot Lake	1	97.8	89.0	32			275
	2	98.7	90.2	95			20
Agnew Lake	1	86.5	91.7	84.5			35
	2	93.0	95.0	93.4			15
Madawaska	1	97.8	89.0	89.8			27
	2	98.2	90.3	93.2			18
Key Lake	1	96.0	-	81.8	70.0	81.0	950
	2	99.5	-	96.2	-	-	200
Rabbit Lake	1	98.0	-	72.0			350
	2	98.7	-	92.8			90

- Indicates not determined.

leaching but through further research this may be decreased to a single stage. Alternatively, chlorine could be interchanged for hydrochloric acid in the first or second stages of the process.

Anhydrous Chlorination

The anyhdrous chlorination route was initially examined on a batch scale in a tube furnace using a 1-inch diameter quartz tube[11]. The bench data indicate that the optimum temperature for uranium recovery is about 400°C while for maximum radium recovery 600°C is required. Hence, a two-stage roast, first at 400°C (allowing for volatilization of U chlorides), then at 600°C, is preferred. Subsequent work was scaled-up to a two-stage counter current vertical shaft furnace, shown in Figure 2[11].

In the continuous operation of the shaft furnace, uranium ore containing 0.12% U and 395 pCi Ra-226/g was chlorinated at a rate of about 750 grams per hour with a retention time of approximately 20 minutes. The calcine was leached in 0.05 M HCl at 80°C for 1 hour, although this acid concentration is not optimized. The results for Elliot Lake are shown in Table 5 indicating good uranium recovery and tails containing 20 pCi/g Ra^{226}. The HCl leach liquor typically contained U 0.05 g/L, Fe 0.3 g/L, Ra^{226} 30,000-50,000 pCi/L.

Similar bench static roast conditions were applied to the complex arsenide-nickel ores of Saskatchewan and followed by the dilute acid leach. Results are also shown in Table 5.

The values for the Saskatchewan ore are preliminary at this time. Optimization tests are required to maximize the uranium recovery and radium solubility. Detailed data are expected late in 1981.

DISCUSSION

The two chloride routes, anhydrous or aqueous, were thus shown to be effective on Elliot Lake ore, for obtaining a high uranium extraction and a tailings containing <25 pCi/g Ra^{226}. Additional work is required on the complex ores to optimize conditions for maximum radium-226 removal. With the relative success of the two individual process routes, a conceptual process flowsheet incorporating both routes can be considered. This is shown in Figure 3.

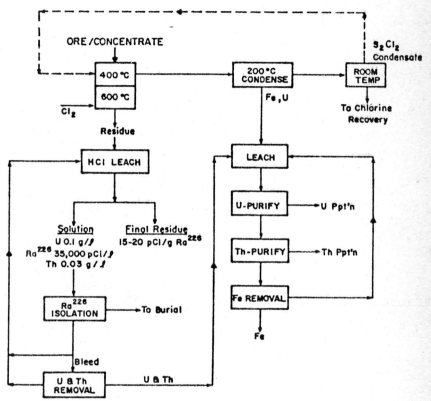

ORE/CONCENTRATE

S_2Cl_2 Condensate

400 °C

600 °C

Cl_2

200 °C CONDENSE

Fe, U

ROOM TEMP

To Chlorine Recovery

Residue

HCl LEACH

LEACH

Solution
U 0.1 g/ℓ
Ra^{226} 35,000 pCi/ℓ
Th 0.03 g/ℓ

Final Residue
15-20 pCi/g Ra^{226}

U-PURIFY → U Ppt'n

Th-PURIFY → Th Ppt'n

Ra^{226} ISOLATION → To Burial

Fe REMOVAL

Fe

Bleed

U & Th REMOVAL   U & Th

FIG. 3

CONCEPTUAL FLOWSHEET

Table 5 Chlorination results

| | Extraction (%) | | | | Residue | |
	U	Ra^{226}	Ni	As	U (%)	Ra^{226}(pCi/g)
Elliot Lake Ore (continuous shaft furnace)	97.6	94.9			0.0029	20
Saskatchewan Ore (bench static tests)	96.3	93.8	84.5	83.6	0.074	410

Essentially, this process is similar to the individual chloride routes, except that slight modifications in the process conditions are made in order to treat concentrates and complex ores. It should be mentioned that the advantage of a preconcentration stage, prior to the chlorination, is to provide for a pyrite concentrate and a uranium concentrate containing also the radio-nuclides. The pyrite concentrate can be subsequently roasted to produce sulphuric acid which is used to regenerate hydrochloric acid for uranium concentrate leaching. Using preconcentration, the leaching plant size is decreased, and in combination with the chloride process there are no sulphides discarded in the tailings. The purification of the off-gases, in the chlorination stage of the process, is far from completed, but studies are in progress. Also isolation of Ra^{226} by sorption, possibly by ion exchange, is underway.

Costs

With respect to economics of processing uranium ores and concentrates through to yellowcake by chloride metallurgy, an economic evaluation of the capital and operating costs have been carried out by an external contractor[12]. The cost studies were on the basis of a daily production of 10,000 lb U$_3$O$_8$/day, treating either low grade ores (2 lb U$_3$O$_8$/ton) or high grade or concentrates (50 lb U$_3$O$_8$/ton). The costs are based on treating the low grade ores by the aqueous chloride route, assuming that anhydrous chlorination would be too expensive. The chlorination route was assumed applicable to the high grade feed materials. The results are given in Table 6. These results, show both processes to be more expensive than the conventional H$_2$SO$_4$ route, although the anhydrous route is probably very close to the cost of conventional processing.

A cost comparison of HCl/Cl$_2$ leaching with conventional H$_2$SO$_4$ leaching has recently been prepared by the authors as shown in Table 7. These costs are for a 5,000 tpd mill fed with an Elliot Lake type ore. Costs are $3.45 or 35% higher for HCl/Cl$_2$ than for H$_2$SO$_4$ leaching. No comparative data are available for high grade (Saskatchewan) ores, but similar cost differences might be expected.

Benefits

It has been demonstrated in the laboratory that the aqueous chloride process will extract uranium, thorium and radium from low grade ores, resulting in 15-25 pCi Ra^{226}/g in the leach residue tails. The process also appears viable for treating high grade ores or concentrates, but further work is required. Because sulphuric acid is required in the process for the regeneration of hydrochloric acid from spent leach liquor, ores containing pyrite can be treated by flotation to remove the pyrite, followed by roasting of the pyrite to produce sulphuric acid on the site. By removing the pyrite, formation of acidic tailings and subsequent long term neutralization of effluents is not necessary. The anhydrous chlorination route, although producing tailings of <25 pCi Ra^{226}/g from low grade ores, still requires considerable optimization to reach such a level with high grade ores or concentrates.

Table 6 Comparative costs of anhydrous and aqueous chloride processes

	Ore grade	
	Low	High
U_3O_8 content of ore, lbs/ton	2.0	50
Ore input, tons/day	5,200	200
U_3O_8 production: lbs/day	10,000	10,000
lbs/year	3,300,000	3,300,000
Process used	Hydrochloric Acid Leach	Chlorination and Hydrochloric acid leach
Direct capital cost $	138,800,000	67,000,000
Direct capital cost, $/annual lb U_3O_8	42.06	20.30
Direct operating cost: $/year	22,540,000	14,050,000
$/ton ore	13.13	213.00
$/lb U_3O_8	6.83	4.26

Table 7   Cost estimates (1980 U.S. Dollars) 5000 tpd plant
0.1% U_3O_8 Elliot Lake ore

	Conventional H_2SO_4 Leach	HCl/Cl_2 Leach
Capital cost	56.6×10^6	115.5×10^6
Direct operating cost	$ 9.96	$ 9.80
Indirect operating costs 25%	2.49	2.45
Interest + Amortization at 10%	6.83	13.94
$U.S./ton ore	$19.28	$26.19
$U.S./lb U_3O_8	$ 9.64	$13.09
$U.S./lb U_3O_8 at 20% interest	$10.80	$15.43

The anhydrous route (uranium) for which the most definitive cost data are available is probably very close to the cost of conventional processing. It should be noted these costs have not included charges or credits for:
- the reduced cost of tailings ponds due to reduced tonnage of tailings (in uranium processing),
- in the case of uranium, the elimination of settling ponds for storage of $BaRaSO_4$,
- long term neutralization of acid effluent is not required, and in the case of uranium, long term treatment of seepage for radionuclide removal is not required,
- by-products recovery is made possible, e.g, sulphur, thorium, arsenic and rare earths,
- the possibility of high purity products at the mill site.

CONCLUSIONS

        The results to date have been sufficiently encouraging to continue
research work on chloride metallurgy. At CANMET we have by no means answered all
of the questions or solved all of the problems. We are however continuing to
apply considerable efforts related to many facets of chloride metallurgy including
anhydrous chlorination, leaching, liquid-solids separation, and solution treatment. Chloride metallurgy offers an alternative to conventional sulphuric acid
leaching through which radium can be quantitatively solubilized. Selective isolation of radium requires further research. We are hopeful that chloride
metallurgy can produce a "walk away" tailings, but of course, at a price.

ACKNOWLEDGEMENTS

The authors gratefully acknowledge the assistance of K.E. Haque, and J.M. Skeaff
in the preparation of this report.

REFERENCES

1   Skeaff, J.M. "Survey of occurrences of Ra^{226} in the Rio Algom Mill, Elliot
    Lake"; Division Report ERP/MSL 79-147 (J); CANMET, Energy, Mines and Resources
    Canada; 1979. Accepted for publication, CIM Bulletin.

2. Ritcey, G.M. "Treatment of radioactive ores at CANMET"; Division Report
    MRP/MSL 77-139 (OP); CANMET, Energy, Mines and Resources Canada; 1977.

3. CANMET Staff - unpublished reports.

4. Skeaff, J.M. "Proposed route for nitric acid leaching of uranium ore and
    removal of Ra^{226} by EDTA"; Internal Report MSL-INT 78-101; CANMET, Energy,
    Mines and Resources Canada; 1979.

5. Ryon, A.D, Hurst, F.J. and Seeley, F.G. "Nitric acid leaching of radium and
    other significant radionuclides from uranium ores and tailings"; Oak Ridge
    National Laboratory, ORNL/TM-5444, 1977.

6. Saint-Martin, N. "Leaching of various radioactive minerals - A literature
    review"; Division Report ERP/MSL 77-89 (LS); CANMET, Energy, Mines and
    Resources Canada; 1977.

7. Honeywell, W.R. and Kaiman, S. CIM Bulletin 59:647:347; 1966.

8. Saint-Martin, N. and Haque, K.E. "Hydrochloric acid leaching from Elliot Lake
    uranium ore - A preliminary study"; Division Report ERP/MSL 77-339 (TR);
    CANMET, Energy, Mines and Resources Canada; 1977.

9. Haque, K.E., Lucas, B.H. and Ritcey, G.M. "Hydrochloric acid leaching of an
    Elliot Lake uranium ore"; CIM Bulletin; 144-147; 1980.

10. Haque, K.E. "Chlorine-assisted leaching of typical Canadian uranium ores";
    Division Report ERP/MSL 80-86 (TR); CANMET, Energy, Mines and Resources
    Canada; 1980.

11. Skeaff, J.M and Laliberte, J.J. "Continuous high temperature chlorination
    of uranium ores"; Division Report ERP/MSL 80-48 (OP); CANMET, Energy, Mines
    and Resources Canada; presented at AICHE Meeting; Portland, Oregon, August
    17-20, 1980.

12. "Evaluation of chlorine-chloride based processes for uranium ores"; Contract
    to Lummus Minerals, DSS No. 15SQ23440-9-9130; November 1980.

REDUCING THE ENVIRONMENTAL IMPACT OF URANIUM TAILINGS BY PHYSICAL
SEGREGATION AND SEPARATE DISPOSAL OF POTENTIALLY HAZARDOUS FRACTIONS

D.M. Levins and R.J. Ring
Australian Atomic Energy Commission, Lucas Heights Research Laboratories
New South Wales, Australia

G.A. Dunlop
Australian Mineral Development Laboratories
Thebarton, South Australia

ABSTRACT

Flotation and hydrocycloning were tested as methods of splitting sulphide
and radionuclide concentrates from the bulk of Australian uranium mill tailings.
Conventional sulphide flotation removed 88-98% of the pyrite in 1-5% of the total
mass. Hydrocycloning was more effective than flotation for concentrating radium
into a low mass fraction. It was found that most of the radium was contained in
the very finest particles (below 5 μm).

A combined flotation/hydrocycloning flowsheet is proposed for segregating
tailings into three fractions for separate disposal. Possible disposal methods
for each of these fractions are discussed.

INTRODUCTION

Uranium mill tailings typically contain almost 75% of the radioactivity originally in the ore, including 2-10% of the uranium and essentially all the thorium-230 and radium-226. The long-term radiological impact of tailings arises because of slow leaching of these radionuclides, continual exhalation of radon, and possible dispersion by wind or water erosion. Tailings also contain heavy metals commonly iron, copper, zinc and lead, and occasionally other toxic elements, such as natural thorium, arsenic, selenium and molybdenum. Uranium is often found in association with sulphides. These can slowly oxidise to sulphuric acid and increase the leaching of heavy metals and radionuclides.

This paper explores the feasibility of physically splitting tailings into a number of fractions according to their radionuclide or sulphide contents. The effectiveness of flotation and hydrocycloning as separation processes is determined for tailings derived from three Australian ores. Options for disposal of the segregated fractions are examined.

DISTRIBUTION OF RADIONUCLIDES IN TAILINGS

Table 1 shows the distribution of radium in North American ore and tailings. While radium is only slightly concentrated in the ore slimes, after leaching, from 84-88% of it reports to the tailings slimes. One possible explanation for this phenomenon is that the debris from the dissolution of uranium minerals is likely to be finely divided and have a high radium content. Another possibility is that the radium redistributes itself during leaching and tends to be adsorbed on the slimes because of their greater surface area. Whatever the reason, any physical process that effectively separates sands from slimes will clearly split the tailings into fractions having a high and low radium content.

Ryon et al. [1] found that tailings contain "hot" grains with a radium content of 50 μCi g^{-1}. Kaiman [2] reported that radium in Elliot Lake tailings is associated with jarosite. This suggested that flotation may be a feasible method of concentrating this mineral [3].

TABLE I. Distribution of Radium between Sand and Slime Fractions of Ore and Tailings [4-7]

	Sands (>75 μm)		Slimes (<75 μm)		^{226}Ra Concentration $\frac{Slimes}{Sands}$
	Weight %	^{226}Ra %	Weight %	^{226}Ra %	
ORE					
Ambrosia Lake	92	82	8	18	2.5
Rio Algom	57	42	43	58	1.8
TAILINGS					
Ambrosia Lake*	80	15	20	85	23
Rio Algom	54	15	46	85	6.6
Denison	49	16	51	84	4.8
Wyoming	69	12	31	88	16
Anaconda	73	12	27	88	17
Beaverlodge	48	16	52	84	4.8

*Sands/slimes split at 150 mesh

We were unable to find any published data on the distribution of thorium-230 in tailings. However Pakkala [6] found that the distribution of natural thorium in Denison tailings closely followed that of radium; 82% of the thorium reported to the slimes, compared with 84% of the radium. Because thorium-230 and natural thorium usually occur in different minerals, it cannot be assumed that natural thorium is an analogue of thorium-230.

SULPHIDES IN TAILINGS

Sulphides in tailings, mainly as iron pyrites, pose an intermediate to long-term problem because of their slow oxidation to sulphuric acid. The rate of oxidation is accelerated in the presence of such autotrophic bacteria as *Thiobacillus thiooxidans, Thiobacillus ferrooxidans* and *Ferrobacillus ferrooxidans*. Given favourable conditions for growth, they may lower the pH to 2-3. Under such conditions, heavy metals in the tailings tend to be mobilised. Bland [8] has estimated that over 95% of the thorium-230 in the top 20 cm of an Elliot Lake tailings pile has been leached in this way. Leaching of thorium-230 from tailings is a long-term problem because, although thorium-230 becomes immobile under neutral conditions, it continues to breed its radiologically more-important daughter, radium. Removal of sulphides from the ore or tailings would remove this possible source of pollution.

SEPARATION PROCESSES

A number of wet classifying techniques are available to split uranium ore or tailings into fractions for separate treatment or disposal. Discussion here will be restricted to flotation, wet high-intensity magnetic separation (WHIMS) and hydrocycloning.

Flotation

This is the commonest process in mineral dressing and is widely used for the recovery of sulphide minerals. Many other minerals can also be separated including heavy-metal oxides, silicates, phosphates, precious metals, carbonates and jarosite. Flotation is most effective on particles within the range 10-800 μm. As particle size is reduced below 10 μm, it becomes increasingly difficult to exploit differences in surface properties to selectively float the required mineral.

Many attempts have been made to beneficiate uranium minerals by flotation but in most cases too high a fraction of the uranium has to be rejected in the tailings for the process to be economic. Successful applications have included; concentration of uranium, gold and pyrite from the Witwatersand ores in South Africa; separation of uranium-bearing copper concentrates in North America, and; production of low-lime concentrates for acid leaching and a low sulphide ore for alkaline leaching at the Beaverlodge mill [9].

Raicevic [3] studied two schemes to separate the pyrite and radioactive minerals from Elliot Lake tailings;

(1) pyrite was floated and the tailings from that process were separately floated for radionuclide recovery, and

(2) pyrite and radioactive minerals were floated together to produce a single concentrate.

Conventional reagents, potassium amyl xanthate (PAX) as collector and pine oil as frother, were used for sulphide flotation. A mixture of fatty acids (mainly oleic) was used to float the radioactive minerals.

Scheme 1 gave the best results. Continuous 50 kg h^{-1} tests on Elliot Lake tailings produced a combined pyrite/radionuclide concentrate containing 98% of the pyrite, 60-65% of the uranium, 73% of the radium and 63% of the thorium in about 25% of the mass. The radium content in the tailings was reduced from 8.5 to 2 Bq g^{-1}. Reagent usage was 35 g t^{-1} PAX, 450-680 g t^{-1} pine oil and 550-1,100 g t^{-1} fatty acids.

Following these encouraging results, plant trials were carried out on a one tonne per hour scale at the Denison mill. Performance was less satisfactory; 98% of the pyrite, 54% of the uranium, 65% of the radium and 75% of the thorium was concentrated in about 30% of the mass. The radium content in the tailings averaged 5 Bq g^{-1}.

WHIMS

Recent research has shown that WHIMS can be used to remove weakly magnetic minerals with particle sizes as low as 15 μm. Extensive testwork in South Africa on reclaimed tailings has shown that about 50% of residual gold and uranium can be recovered in about 10% of the feed [10]. A limitation on the use of WHIMS is a drop in efficiency of collection for particles below 20 μm.

Raicevic [11] has proposed a process that uses WHIMS and flotation to beneficiate uranium and remove pyrite from uranium ore. Pyrite is first removed from the ground ore in a conventional flotation circuit. The ore is then deslimed and both sands and slimes are subjected to WHIMS. The non-magnetic fraction of each is returned to the mine as backfill. Laboratory experiments with Elliot Lake ore have produced a material suitable for backfill containing <1 Bq ^{226}Ra g^{-1}. About 98.5% of the pyrite, 97% of the uranium, 98.5% of the radium and 92% of the thorium can be concentrated into a weight of about one-third that of the original ore. If these results can be reproduced for a variety of ores under plant conditions, WHIMS could significantly reduce the volume of tailings for disposal.

Hydrocyclones

Since about 85% of the radium is contained in the minus 75 μm fraction, a simple classifying process involving hydrocyclones may be effective in concentrating most of the radionuclides into a small fraction of the total mass.

Hydrocyclones are widely used in uranium milling; applications include sizing in closed-circuit ball milling, desliming, and counter-current washing of sands. The advantages of hydrocyclones over other solid/liquid separation methods are high capacity, low initial cost and predictable operation.

The cut size (d_{50}) of a cyclone is defined as the particle diameter, half of which reports to both underflow and overflow. The overall efficiency of separation depends on the particle size distribution of the feed and the performance curve of point efficiency versus particle diameter. Figure 1 gives typical point efficiency curves for 75-600 mm diameter cyclones operating at a slurry density of 25 weight % [12]. Cut diameter increases from 10-35 μm as cyclone diameter is increased.

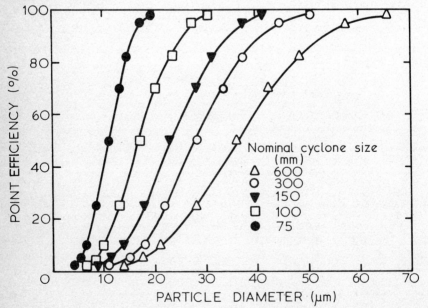

Figure 1 Typical Point Efficiency Curves for Hydrocyclones

In order to concentrate radionuclides into as small a mass as possible, a small cut size may be desirable. This can be achieved with a small diameter cyclone but pressure drop is increased and capacity is reduced. For a slurry feed density of 25 weight %, capacities range from about 2 t h^{-1} (solids) for a 75 mm diameter cyclone to 100 t h^{-1} for a 600 mm unit. In applications where a high throughput and very fine size separation are needed, a large bank of parallel cyclones would be required.

EXPERIMENTAL STUDY

An experimental study, undertaken at the Australian Mineral Development Laboratories (AMDEL), was originally aimed at determining the feasibility of separating sulphides and radionuclides from Australian tailings using the flotation process proposed by Raicevic [3]. Later, the possibility of simplifying the process was examined. Following preliminary flotation and classification tests, an improved process involving cycloning and flotation was developed.

Table II shows the radionuclide and sulphide contents of the three Australian tailings chosen for study. None had particularly high thorium or sulphide levels, although they were generally sufficient to track their fate through the flotation and hydrocycloning circuits.

Table II. Radium, Thorium, Uranium and Sulphide Contents of Tailings

Element	Tailings A	Tailings B	Tailings C
^{226}Ra (Bq g^{-1})	43	7.8	167
Th (%)	0.002	0.020	<0.002
U (%)	0.021	0.0095	0.058
Sulphides (% S)	0.37	0.68	0.01

Radionuclide Distribution

The size distribution of each tailings sample was determined by wet screening down to 350 mesh (45 μm) and by a Warman "cyclosizer" for the finer fractions. The minus 5 μm fraction was separated from the final cyclone overflow by filtration. Figures 2-4 show the distribution of mass and radionuclides in each tailings sample.

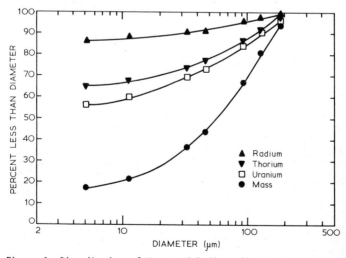

Figure 2   Distribution of Mass and Radionuclides for Tailings A.

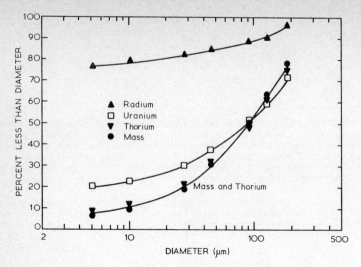

Figure 3 Distribution of Mass and Radionuclides for Tailings B.

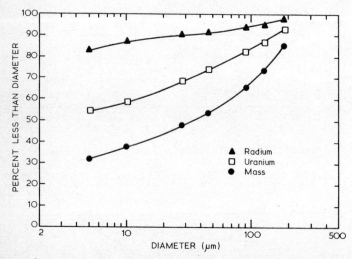

Figure 4 Distribution of Mass and Radionuclides for Tailings C.

The distribution of radium in these Australian tailings was similar to that previously reported for North American tailings [4-7]. The percentage radium in the slimes (<75 μm) ranged from 87-94%. Sub-sieve analysis further revealed that most of this radium was contained in the minus 5 μm fraction. Unleached uranium also tended to concentrate in the slimes though not to the same extent.

Thorium distribution was determined for two tailings only; it tended to con-centrate in the slime fraction of Tailings A but was almost uniformly distributed in Tailings B. This dissimilar behaviour could have been caused by different thorium mineralisation in the two tailings.

Flotation of Sulphides and Radionuclides

All flotation tests were done in an Agitair LA-500 machine with an impeller speed of 17 rev s^{-1}. Cell sizes were selected to suit the various flotation steps and to provide a pulp density of 30-32 weight % in the roughing and scavenging stages. Pine oil and methyl isobutyl carbinol (MIBC) were tested as frothers. PAX was used as the collector for sulphides, while single distilled oleic acid (SDOA), used by Raicevic [3], and an Australian collector (71% oleic acid/29% linseed fatty acid - OA/LFA) were compared for radionuclide flotation. Use of either SDOA or OA/LFA gave similar results.

Initial flotation tests using Tailings A were carried out under the conditions recommended by Raicevic [3]. Reagent additions were not optimised. Figure 5 shows the flowsheet and reagent additions for the flotation circuits. Typical results are summarised in Table III. Sulphur recovery was only 88% but later work showed the unrecovered sulphur was present mainly as sulphate rather than sulphide. Recovery of radionuclides was comparable with Raicevic's results but the weight of uncleaned concentrate was unacceptable by plant practice. At least 50% of the concentrate would have to be rejected in cleaning and this would result in significant radionuclide losses. The radium content of the tailings was reduced to half its value in the feed. This was not sufficiently encouraging to justify further work on flotation of radionuclides.

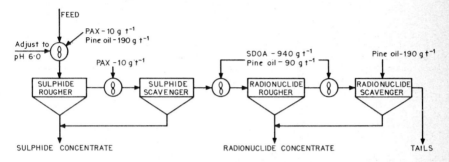

Figure 5   Flowsheet for Flotation of Sulphides and Radionuclides from Tailings.

TABLE III. Removal of Sulphides and Radionuclides from Tailings A by Flotation

Product	Distribution (%)				
	Mass	S	U	Th	Ra
Sulphide concentrate	12.8	81.2	14.3	18.7	10.6
Radionuclide concentrate	33.1	7.1	42.3	47.5	61.6
Combined concentrate	45.9	88.3	56.6	66.2	72.2
Final Tailings	54.1	11.7	43.4	33.8	27.8

Hydrocycloning

Particle size analysis of Tailings A showed that 89% of the radium was contained in the minus 20 μm fraction (see Figure 2). Cycloning, therefore, appeared to be a suitable alternative to flotation for isolation of a radium-rich fraction.

Tests were carried out with two 75 mm Warman cyclones operating in series. The underflow from the first cyclone passed to a second cyclone. The cut size (d_{50}) was estimated to be about 15 μm. Table IV shows that the split obtained using a double pass cyclone circuit was only slightly inferior to that achievable with close particle sizing. Over 85% of the radium was separated into a stream containing less than 25% of the total weight. Thorium and uranium were also enriched in the overflow but not to the same extent as radium. Sulphides were not effectively separated by cycloning.

TABLE IV. Classification of Tailings A by Hydrocycloning

Product	Distribution (%)				
	Mass	S	U	Th	Ra
Underflow	76	83.1	54.3	43.0	13.9
Overflow	24	16.9	45.7	57.0	86.1

Combined Hydrocycloning and Flotation

Following encouraging results with hydrocycloning, the effectiveness of a combined hydrocycloning and flotation circuit was determined. For maximum information the tailings were split into three fractions, a sands (nominally >45 μm), a fines (nominally 15-45 μm) and a slimes (nominally <15 μm). Sulphides in each of these fractions were then separately floated.

The flowsheet is shown in Figure 6. The initial split at 45 μm was done by hand wet screening and, as such, was imperfect which reflects the inefficiency in practice of cyloning to 45 μm. In some runs the effectiveness of slimes cleaner and scavenger circuits was studied but these had little effect on the overall material balances.

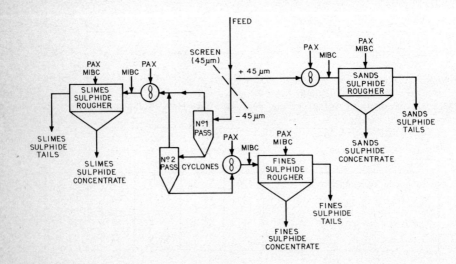

Figure 6. Experimental Scheme for the Flotation and Cycloning of Tailings

Results are tabulated in Table V. Pyrite flotation was very efficient, 88-98% of the sulphides was concentrated in 1-5% of the total mass. Essentially all the unrecovered pyrite was contained in the slimes which proved difficult to float. Radionuclide recovery in the slime and fine fractions was generally good and closely followed that expected from size analysis. Specific findings for each tailings are as follows.

Tailings A. Most of the radionuclides, including 83% of the radium, were concentrated in only one-sixth of the total mass. This was far superior to flotation where only 62% of the radium was concentrated in one-third of the total mass.

Tailings B. For these tailings, the best approach would be to aim for a split at about 45 μm (i.e. combine fines and slimes). About 80% of the radium would then be concentrated in one-quarter of the mass. The radium content of the sands is only 1.25 Bq g^{-1} which is not significantly above the maximum allowable content of radium in building materials in the UK of 0.9 Bq g^{-1} [13]. As expected from the size distribution (Figure 3), thorium and uranium do not concentrate significantly in the slimes.

Tailings C. This gave the poorest result; only 67% of the radium and less than half the uranium and thorium was concentrated in the slimes which was 31% of the total mass. From size distribution data (Figure 4), up to 83% of the radium should be concentrated in this size fraction. Further work is in progress with the aim of obtaining a better split. However, because of the high clay content (30% below 5 µm) of this ore, the radium cannot be isolated in a small mass fraction. The sulphide distribution in the tailings is given for completeness only; in practice the flotation step would be unnecessary because of the low sulphide content of the tailings.

TABLE V. Performance of Cycloning/Flotation Process

Tailings	Fraction	Distribution (%)				
		Mass	Sulphides	U	Th	Ra
A	Sulphide concentrate					
	Sands	2.8	59.0	2.1	2.1	0.3
	Fines	2.1	37.2	1.9	3.7	0.4
	Slimes	0.3	1.6	0.7	0.8	0.4
	Total	5.2	97.8	4.7	6.6	1.1
	Tails					
	Sands	58.7	<0.1	27.7	14.8	7.9
	Fines	19.6	0.3	16.2	16.1	7.8
	Slimes	16.5	1.9	51.4	62.5	83.2
B	Sulphide concentrate					
	Sands	2.3	68	4.0	1.3	0.8
	Fines	0.7	26	1.5	0.4	1.2
	Slimes	0.4	0.5	1.5	0.7	5.8
	Total	3.4	94.5	7.0	2.4	7.8
	Tails					
	Sands	71.9	<0.1	62.2	71.6	11.8
	Fines	19.4	<0.1	13.6	17.4	9.4
	Slimes	5.3	5.5	17.2	8.6	71.0
C	Sulphide concentrate					
	Sands	0.6	42	0.6	1.2	0.6
	Fines	0.2	42	0.3	0.6	0.3
	Slimes	0.7	4	1.4	1.6	1.9
	Total	1.5	88	2.3	3.4	2.8
	Tails					
	Sands	50.0	<0.1	29.2	17.0	10.7
	Fines	18.0	<0.1	21.7	34.7	19.7
	Slimes	30.5	12	46.8	44.9	66.8

PROCESS AND COST CONSIDERATIONS

Figure 7 shows a schematic flow diagram for a process to remove sulphides from tailings and concentrate radionuclides in a relatively small mass. Flotation of sulphides is proposed as the initial step followed by cycloning. A two-pass cyclone circuit is desirable because even a small percentage of slimes in the underflow will have a detrimental effect on the decontamination of the sands fraction. The cut size of the cyclone should be determined from a knowledge of the particle size and radionuclides distribution, and the proposed disposal plan.

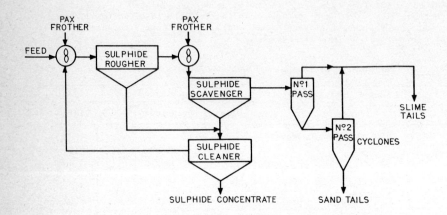

Figure 7   Proposed Flowsheet for the Separation of Sulphides and Radionuclides from Tailings

The cost of this process will be much lower than any schemes involving chemical decontamination. Capital costs for sulphide flotation equipment are estimated to be $40-120 (1980 US dollars) for each tonne of solids processed per day [14]. Power costs for flotation are typically 2-3 kWh per tonne, while operating costs should range from $0.30 - $1.00 per tonne. Reagent costs account for about half the total operating cost [14]. The cost of cycloning is also low; the major capital expense being for pumps and piping. Power costs are estimated to be between 0.1-0.4 kWh per tonne solids depending on the cut diameter of the cyclones. A small cut diameter (<30 μm) will require higher capital and power costs than a more conventional cut diameter (40-70 μm). The overall cost of the process shown in Figure 7 is likely to range from $0.50 to $1.50 per tonne tailings.

DISPOSAL OPTIONS

Having demonstrated separation of the tailings into a sulphide concentrate, a low activity sands and a slime fraction containing most of the radionuclides, we consider now the options available for further treatment or disposal of these fractions.

Sulphide Concentrate

About 80% of the world's uranium mills leach the ore in sulphuric acid. An obvious way of disposing of the sulphide concentrate is to convert the sulphides to sulphuric acid. This would reduce reagent costs in the mill. An ore containing 1.5% sulphides is theoretically capable of supplying all the sulphuric acid requirements (45 kg t^{-1}) for the model mill [15]. If the sulphide content of the ore is below the equivalent acid consumption of the ore, it may be possible to process a small quantity of high sulphide waste rock to make up the difference.

The sulphide concentrate can be oxidised by roasting, bacterial oxidation or chemical oxidation at elevated temperatures. Roasting is the most conventional route but, as the sulphide concentrate has a fairly high radionuclide content (see Table V), airborne pollution could result. Bacterial oxidation is also feasible but is slow, and the relatively weak acidic solutions produced could pose control problems in the ore leaching circuit. Wet chemical oxidation at 150-185°C using air as oxidant appears to be an attractive route since strong acid solutions (>50 g L^{-1}) can be generated within a few hours [16]. The heat of reaction of the sulphides should be capable of sustaining the required reaction temperature under most conditions. The acidic solutions produced will have a high iron concentration which is often an advantage in leaching ores [17].

If the production of sulphuric acid from pyrite is not attractive, the sulphide fraction could be buried away from the tailings preferably in an alkaline environment.

Sands

Where uranium is mined underground (about 50% of world-wide production), the most suitable disposal method for the sands fraction is as mine backfill. This enables pillars to be mined and uranium resources to be increased by about 25-30% [18].

Fill for mines is normally cycloned to keep the minus 10 μm fraction below 10 per cent [19]. This improves dewatering rates and increases the strength of the fill. Cement (3-10%) is sometimes added to increase the strength further and to provide a better surface for manoeuvring of mining equipment.

Backfilling of uranium mines has not been practised extensively in the past although there has been an upsurge in interest in recent years. The main objection has been the increase in working-level in the mine as a result of increased radon emanation. However, the present work has shown that, if sands and slimes are well separated by two passes through cyclones, the radium content (and hence the radon emanation rate) will be low.

In open-cut mining, the separated sand fraction is likely to be stored in a conventional tailings dam. Because of its relatively low radionuclide content, rehabilitation requirements are likely to be less stringent. For example, above-ground disposal may be acceptable in lieu of returning the sands to the pit. In some cases, when low grade ores are processed, it may be possible to reach 'de-minimus' levels by cycloning alone.

Slimes

The radionuclide-rich slime fraction poses the most difficult disposal problem but its relatively small mass enables more expensive solutions to be considered. If the slimes alone are stored in a tailings dam, the stability of the structure must be carefully considered. One approach is to eliminate the tailings dam altogether by dewatering on a belt filter, granulating the slimes and disposing of them in the mine or in specially excavated trenches.

Granulation involves tumbling of moist powders to produce near spherical pellets normally with a diameter of 3-15 mm. Since the product from belt filtration usually contains 20-25 weight per cent moisture, it may be necessary to add a small quantity of cement or other dry material to obtain satisfactory granulator performance. Although the pellets are in an ideal form for handling and disposal, they do not have a high strength and will ultimately disintegrate. Tailings pellets that are fired to form sintered particles are much stronger and highly resistant to leaching of radium [20].

The cost of granulation on a 100-300 t h^{-1} scale is estimated to be about US$0.40 per tonne if no binder is used, about US$0.80 per tonne with cement added, and US$8 per tonne if the pellets are fired [21]. As only a small fraction of the tailings need be granulated, the cost of even the most expensive option is not prohibitive.

Other approaches to slimes treatment or disposal include electrokinetic consolidation [22], incorporation in concrete, and chemical leaching to remove the radionuclides [1]. There is a clear need for further research into methods of slime disposal.

CONCLUSIONS

.   Most of the radium in uranium mill tailings is concentrated in the very finest particles (<5 μm).

.   Other radioactive elements, thorium and uranium also tend to concentrate in the slimes, although not to the same extent as radium.

.   Conventional sulphide flotation of Australian tailings removed 88-98% of the pyrite in 1-5% of the mass.

.   Hydrocycloning was more effective than flotation for concentrating the radium fraction of Australian tailings. In one experiment, 83% of the radium was concentrated in one-sixth of the total mass using a two pass cyclone circuit, whereas with flotation only 62% of the radium was concentrated into one-third of the mass.

.   Possible options for disposal of the separated fractions are wet oxidation of the sulphide concentrate to produce sulphuric acid for use in the mill, backfilling of the sand fraction, and granulating of the slimes followed by disposal in specially excavated trenches.

REFERENCES

1. Ryon, A.D., Hurst, F.J. and Seeley, F.G. : "Nitric Acid Leaching of Radium and Other Significant Radionuclides from Uranium Ores and Tailings", ORNL/TM-5944, 1977.

2. Kaiman, S. : "Mineralogical Examination of Current Elliot Lake Uranium Tailings", Division Report MRP/MSL 77-340(J); CANMET, Energy Mines and Resources Canada, 1977.

3. Raicevic, D. : "Decontamination of Elliot Lake Uranium Tailings", CIM Bulletin 72 (808), (1979).

4. Seeley, F.G. : "Problems in the Separation of Radium from Uranium Ore Tailings", Hydrometallurgy 2, 249-263 (1976/77).

5. Skeaff, J.M. : Survey of the Occurrence of Ra-226 in the Rio Algom Quirke I Uranium Mill, Elliot Lake, CIM Bulletin 74 (830), 115-121 (1981).

6. Pakkala, T.E. : Denison Mines Limited, Private Communication

7. Lakshmanan, V.I. and Ashbrook, A.W. : "Radium Balance Studies at the Beaverlodge Mill of Eldorado Nuclear Limited", Proc. OECD-NEA Seminar on Management, Stabilisation and Environmental Impact of Uranium Tailings, pp. 51-64, Albuquerque, July 24-28, 1978.

8. Bland, C.J. : "A Method of Computing the Rate of Leaching of Radionuclides from Abandoned Uranium Mine Tailings", 9th Annual Hydrometallurgical Meeting of the Canadian Institute of Mining and Metallurgy - Uranium Processing and the Environment, Toronto, Nov 11-13, 1979.

9. Simonsen, H.A., Boydell, D.W. and James, H.E. : "The Impact of New Technology on the Economics of Uranium Production from Low-Grade Ores", Fifth Annual Symposium of the Uranium Institute, London, Sept 2-4, 1980.

10. Corrans, I.J. and Levin, J. : "Wet High-Intensity Magnetic Separation for the Concentration of Witwatersand Gold - Uranium Ores and Residues", J. South African Inst. Min. Met. 79 (8), 210-228 (1979)

11. Raicevic, D. " "Removal of Radionuclides from Uranium Ores and Tailings to Yield Environmentally Acceptable Waste", First International Conference on Uranium Mine Waste Disposal Chap. 25, pp. 351-360, Vancouver, May 19-21, 1980.

12. Zanker, Z. : "Hydrocyclones : Dimensions and Performance" Chem. Engng. 84 (10), 122-125 (1977).

13. Riordan, M.C., Duggan, M.J., Rose, W.B. and Bradford, G.F. : "The Radiological Implications of Using By-Product Gypsum as a Building Material", National Radiological Protection Board, England, 1972.

14. Perry, R.H. and Chilton, C.H. (Eds) : Chem. Engrs. Handbook, 5th Edition pp. 21-67 to 21-68, McGraw-Hill, New York.

15. U.S. Nuclear Regulatory Commission : Final Generic Environmental Impact Statement on Uranium Milling, NUREG-0706, 1980.

16. Johnson, P.H. : "Acid-Ferric Sulphate Solutions for Chemical Mining", Min. Sci. Eng. 6 (2), 64-68 (1965).

17. Ring, R.J. : "Ferric Sulphate Leaching of Some Australian Uranium Ores", Hydrometallurgy 6, 89-101 (1980).

18. Watts, Griffis and McOuat Limited : "Study of the Use of Tailings as Backfill in Uranium Mines", Watts, Griffis and McOuat Limited, Consulting Geologists and Engineers, Toronto, Ontario, 1978.

19. Thomas, L.J. : An Introduction to Mining. Mentheun, Sydney, 1978.

20. Wiles, D.R. : "The Leaching of Radium from Beaverlodge Tailings", Proc. OECD-NEA Seminar on Management, Stabilisation and Environmental Impact of Uranium Tailings, pp. 245-258, Albuquerque, July 20-28, 1978.

21. Hall, J. : Commonwealth Scientific Industrial Research Organisation (CSIRO), Division of Mineral Engng. Victoria Australia, Private Communication

22. Sprute, R.H. and Kelsh, D.J. : "Electrokinetic Consolidation of Slurries in an Underground Mine", USBM RI-8190, 1976.

REDUCTION OF RADIONUCLIDE LEVELS IN URANIUM MINE TAILINGS

M.H.I. Baird, S. Banerjee[1], A. Corsini, D. Keller,
S. Muthuswami, I. Nirdosh[2], A. Nixon, A. Pidruczny,
M. Tsezos, S. Vijayan[3] and D.R. Woods

McMaster University
Hamilton, Ontario, Canada

ABSTRACT

This paper considers several alternatives to conventional acid-leach uranium milling as applied to ores from the Elliot Lake area of Ontario, Canada. The objective is to discharge bulk tailings containing no more than 20 pCi of ^{226}Ra per g of solids. Experimental work at the laboratory scale has been carried out on a flotation separation of the ore, ferric chloride leaching of both U and Ra from ore, and leaching of radium from acid-leached solids by means of a chelating agent (EDTA). Preliminary work has also been carried out on radium removal from solution by biosorption. The results of these several studies are promising but in need of economic and engineering assessment.

1. S. Banerjee is now with the Department of Nuclear Engineering, University of California, Santa Barbara, CA.

2. I. Nirdosh is now with Lakehead University, Thunder Bay, Ontario.

3. S. Vijayan is now with Atomic Energy of Canada Ltd., Pinawa, Manitoba.

1. Introduction

The question of radionuclides in uranium mine tailings has received increasing attention in the past ten years, due to an increased level of public concern and to the prospect that uranium production will continue to increase in the foreseeable future. Canada is one of the world's largest uranium producers; production is currently about 7000 t/y and is expected [1] to rise to 11000 t/a by the year 1990. The large deposits of uranium in Ontario assay about 0.1% uranium, so millions of tons of tailings are discharged each year in this area alone. The other main uranium mining area in Canada is northern Saskatchewan, where the ore grades are higher, in the range 0.2% up to 5% or more. This results in a smaller relative volume of tailings per tonne of uranium extracted, but the radionuclide concentration is higher.

A major part of the attention given to uranium tailings has been studies of existing tailings areas, and the means of preventing radionuclides from entering the environment. Proposed new uranium mills and extensions to existing mills are subject to exhaustive environmental assessments in this respect [2,3] to ensure compliance with federal and provincial standards. The conventional wisdom is that radionuclides should as far as possible be "contained" in the solid tailings behind a dam, with suitable chemical treatment (e.g. liming to avoid acid conditions) or top vegetation to prevent dusting and reduce radon emission. Aqueous effluents can be treated effectively by various means, principally barium radium sulphate coprecipitation which has been practiced successfully for the past 20 years.

The half-life of ^{226}Ra is 1600 years and that of its precursor ^{230}Th is 75,000 years. This raises difficult technical questions about the possible spread of radionuclides from mine tailings over very long times; it also raises economic and ethical questions as to whether we could expect effluent water from tailings areas to be treated long after a uranium mine and mill has ceased production.

Such considerations have led some researchers to an alternative approach, namely to investigate deliberate extraction of radionuclides from tailings. These investigations have been reviewed recently [4,5,6] and in summary, it may be said that

(i)   The only radionuclide considered in detail is ^{226}Ra.

(ii)   Radium can be extracted to below the acceptable level in the solids (20 pCi/g) only with the use of reagents which are corrosive (HNO_3) or unduly expensive.

(iii) The techniques involved are mainly chemical. A promising exception is mineral flotation [7].

A third option can be considered, in addition to those of containment and extraction of the tailings. This is the modification of the milling process so that radionuclides can be recovered as a side stream in a concentrated form which can easily be removed for safe long-term storage, leaving the millions of tons of tailings with radionuclide levels that are considered acceptable. This option appears to be of relatively minor research interest so far; thus, a recent 155-page survey on radionuclide removal from process streams [4] devotes only 12 pages to alternate processing technology. Another extensive recent Canadian survey [5] concludes that "research into a new or modified uranium recovery process is of high priority" with the dual objectives of radionuclide separation and improved uranium recovery economics. It may be argued against the third option that it does nothing for existing tailings areas, but the future projections of uranium production greatly exceed the total of past production. In some cases it might be possible to bury "old" tailings under "new" tailings which have acceptable activity due to the use of a suitably modified milling process.

This presentation gives an overview of results obtained so far in a group project at McMaster University funded by a strategic grant from Canada's Natural Science and Engineering Research Council. The interdisciplinary group was formed in 1979 with five research staff and several faculty members from the departments of Chemistry, Chemical Engineering and Engineering Physics, under the leadership of S. Banerjee. The group's broad objective has been to examine possible processes for isolating radium (and ^{230}Th) in a concentrated form for disposal. As the project has developed, emphasis has shifted increasingly towards modifications of the existing uranium milling process, rather than treatment of tailings. Thus far, ^{230}Th has not been considered. It is, however, important to determine ^{230}Th as distinct from the much more abundant but less radioactive ^{232}Th.

Several different possible process modifications are currently under study, and are discussed below in the order in which they occur in the basic milling process. The first modification is flotation (Sect. 3 below) as a means of concentrating U and Ra to leave the bulk of the tailings with acceptably low radionuclide levels. The second approach (Sect. 4 below) is to leach both U and Ra from the ore. Sect. 5 describes a method of leaching Ra from U-leached solids before they are discharged from the mill. Studies on a new method of recovery of Ra from solution are described in Sect. 6. Economic analysis will eventually determine whether any of these modifications are feasible. In the meantime, research results are being gathered in the context of integrated processes, i.e. using recycle concepts with reagent recovery rather than merely "once-through" treatments. The results are not yet at the stage where they can be published in full, and this presentation must therefore be taken as a summarised progress report. The names of the principal investigators are given in parentheses in the headings of the appropriate sections.

2.   Conventional Milling Process at Elliot Lake

The attention of the McMaster group has so far been confined to milling processes and ores characteristic of the Elliot Lake area of Ontario. The ores found in this region are hard and granite-based, with uranium occurring to the extent of about 0.1%, mainly as uraninite, $(U,Th)O_2$ and brannerite, $(U,Th)Ti_2O_6$. The radium level in the ore is about 350 pCi/g. A significant amount of pyrite (FeS_2) is present, typically in the order of 5%.

Because the ore contains only slight amounts of "acid consumers" such as calcite, a sulphuric acid leaching process is employed. There are several mines and mills in the area, operated by Rio Algom Ltd. and Denison Mines Ltd.; recent descriptions of the respective milling processes are available [8,9] and the process at the Quirke Mill of Rio Algom Ltd. is shown as an example in Figure 1.

The ore is finely ground (50% below 74 μm) and leached with sulphuric acid for a contact time of 40 h at about 70°C, in banks of air-blown pachuca tanks. It is important to have oxidising conditions in order that at least some of the iron in solution is maintained in the ferric form, permitting oxidation of quadrivalent uranium to the soluble hexavalent form.

$$UO_2 + 2 Fe^{3+} -> UO_2^{2+} + 2 Fe^{2+}$$

The slurry discharged from the pachucas contains about 50 g free sulphuric acid per litre of liquid, with about 96% of the uranium dissolved as uranyl sulphate. The pH of this slurry is raised to 1.8 by addition of finely ground limestone or slaked lime and the uranium-bearing solution (pregnant liquor) is then separated from the solids in an arrangement of cyclones and countercurrently operated thickeners or, alternately, with a vacuum drum filter. The pregnant liquor is sent to ion exchange circuits for uranium concentration, and eventual formation of yellowcake.

Since ^{226}Ra has an extremely insoluble sulphate, it is almost entirely retained in the solid tailings [10]. Skeaff [11] and other previous workers have reported that the radium concentrations are greatest in the fine size fractions, indicating that dissolution and surface redeposition occur at some stage in the leaching. Chemically, the radium concentration in (or on) the solids is minute compared with those of $CaSO_4$ and $Fe(OH)_3$ formed in the partial neutralization. Most of the pyrite remains undissolved in the leaching process, posing an additional environmental problem in the tailings as it is very slowly oxidised by air (catalysed by bacteria) in moist conditions to form sulphuric acid.

3.   Flotation Treatment of Ore (S. Muthuswami, S. Vijayan, D.R. Woods)

To overcome the problem of leaving pyrite and radionuclides in large quantities of the waste material, this part of the project focuses on separating

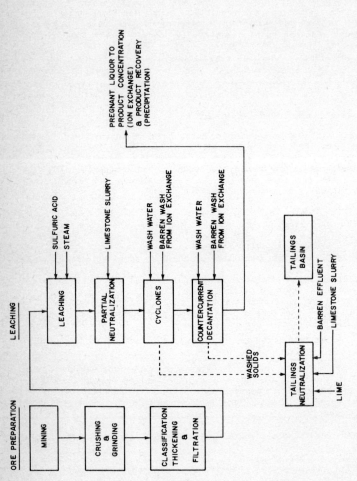

Figure 1 Milling process at Quirke Mill (Rio Algom Ltd.), Elliot Lake Ontario.

ORE PREPARATION

MINING → CRUSHING & GRINDING → CLASSIFICATION THICKENING & FILTRATION

LEACHING

LEACHING → PARTIAL NEUTRALIZATION → CYCLONES → COUNTERCURRENT DECANTATION

SULFURIC ACID
STEAM

LIMESTONE SLURRY

WASH WATER
BARREN WASH FROM ION EXCHANGE

WASH WATER
BARREN WASH FROM ION EXCHANGE

PREGNANT LIQUOR TO PRODUCT CONCENTRATION (ION EXCHANGE) & PRODUCT RECOVERY (PRECIPITATION)

WASHED SOLIDS

TAILINGS NEUTRALIZATION

LIME

BARREN EFFLUENT
LIMESTONE SLURRY

TAILINGS BASIN

- 176 -

these substances along with uranium minerals by flotation as a pyrite concentrate and a uranium concentrate, leaving the bulk of the mass of the ore as harmless flotation tailings. The two concentrates could be treated according to the method described in Section 4.

Based on the ore assay and on the requirement of less than 20 pCi Ra/g of tails, simple mass balance calculations show that this would require a separation scheme that would separate 97% of the pyrite and 97% of the uranium, in about 25% of the mass of feed ore, with the radionuclides concentrating wherever uranium concentrates. Work by Raicevic [7] and others suggested that physical separation techniques such as magnetic separation and flotation might be technically capable of achieving this target objective.

Previous Flotation Work

Successful flotation requires a careful choice of the collector, gangue depressant, and frother. For uranium flotation, it has been found that the collectors oleic and myristic acids, petroleum sulfonates, and iso-octyl phosphoric acid [12-15] floated uranium but not to within our target range. Some of the poor results occur because for acid collectors, polyvalent metal ions such as Ca^{2+}, and Fe^{3+} activate the silica even when depressants are added so that poor separation occurs [16,17]. Iso-octyl phosphoric acid is selective but the uranium recovery does not satisfy the target requirements [12,18]. Marabini and coworkers [19] floated uranium mineral from a mixture of pitchblende and quartz at pH 1.0 in a Hallimond tube with 95% uranium recovery with a grade of 98% uranium. Higher recoveries of uranium were obtained at the cost of lowering the grade. They used a mixture of cupferron (a chelating agent), hydroxylamine (reducing agent), and fuel oil (promotor) in this flotation.

Scouting Tests

Four acids, two amines, two high molecular weight chelating agents and two low molecular weight chelating agents were used as collectors in uranium flotation from Elliot Lake ore (see Table I). The reagents selected had good affinity for uranium.

Three experimental methods were used in this work to float uranium mineral.

Method 1: Uranium float with no Pyrite Prefloat.

Elliot Lake ore from Quirke Mill (Rio Algom Ltd.), was ground to minus 100 mesh and screened to remove the minus 400 mesh fraction. This step was taken to avoid any slime problem in the flotation. From this ground feed, 200 g was mixed with 1.78 L distilled water and appropriate quantities of the uranium collector were added. Uranium was floated in a bulk flotation, using an Agitair flotation cell.

Method 2: Pyrite Prefloat, Delayed Uranium Float.

A feed of 200 g ore (prepared as given in Method 1 (minus 100 plus 400 mesh)) was mixed with 1.78 L distilled water, and pyrite was floated with potassium amyl xanthate as collector at a dosage of 40 g/t (0.08 lb/t) of dry feed, and Dowfroth 250 as frother at a dosage of 25 g/t (0.05 lb/t). The pyrite-free tailings from several such flotations were mixed, air dried and kept as stock feed for uranium flotation. Later, 200 g portions of this pyrite-free ore were floated for uranium in an Agitair flotation cell as described in Method 1.

Method 3: Pyrite Prefloat, Immediate Uranium Float

A sample of 500 g of ground ore was taken from the grinding circuit at the Quirke Mill, and this sample (50% minus 400 mesh) was further ground in the Process Development Laboratory near the mill to 85% minus 400 mesh; pyrite was floated as given in Method 2.

Immediately uranium was floated in the same pulp with a chosen collector. Tap water was used and a Denver cell was used in this flotation. The success of this method indicated that fine particles do not present any problem.

Analysis

Uranium was analysed in the float, tail and in the solution; radium was assayed in selected tests; thorium-230 was not analysed in any test.

Results

Some of the results obtained are shown in Table I. The results with amine collectors were poor. Both the uranium recovery and the uranium grade in the float were low. Adogen was used as saturated aqueous solution at pH 1.5, and laurylamine at a concentration of 0.1 mmol/L. The solubility of Adogen in water is very low.

Table I

Results of Scouting Tests

Reagent	Chemical Nature of reagent	Float Method	% Distribution					
			Pyrite Float		Uranium Float		Uranium Tails	
			mass	U	mass	U	mass	U
Acids								
Procol CA540	sulfosuccinate	2	16	40	15	34	69	26
Aerosol OT	sulfosuccinate	1	–	–	40	56	60	44
Petroleum sulfonate		1	–	–	69	90	31	10
Neofat 94-04	fatty acids[a]	2	16	40	26	44	58	16
Amines								
Adogen 364	tertiary amine	1	–	–	26	37	74	63
Lauryl amine	primary amine	2	16	40	31	16	53	44
Chelating Agents High MW								
TBP	tributyl-phosphate	2	16	40	8	25	76	35
Kelex 100	oxime derivative	2	16	40	3	13	81	47
Low MW								
Cupferron		3	7	9	25	83	68	2[b]
Salicylic acid		3	7	13	14	58	79	19[c]

a   major components: oleic acid 76%; linoleic acid 7%, palmitoleic acid 7%.

b   about 6% U is in solution

c   about 10% U is in solution.

Acidic collectors, Aerosol OT and petroleum sulfonate, floated both uranium mineral and silica indiscriminately.

The other two acidic collectors, Neofat 94-04 and Procol CA 540, were slightly better than the above two, but the results did not meet our goal. Both the recovery and the grade of uranium were poor.

Chelating collectors gave encouraging results, especially cupferron. With cupferron it was possible to achieve 97% uranium (including 7% U in solution) and radium recovery in about 30-35% mass. The flotation tailings assayed 0.03-0.08% sulphur and 10-15% pCi Ra/g. The improvement of flotation with cupferron, with Kelex 100, and scouting tests with other chelating agents are under way.

Proposed Flow Sheet

A conceptual flow sheet for Elliot Lake ore treatment that includes flotation separation is given in Figure 2, based upon the test results described above.

4.   Inorganic Salt Leachants for Ore (I. Nirdosh, M.H.I. Baird)

The initial objective of this phase of the project was to compare the effectiveness of various inorganic leachants for combined extraction of uranium and radium from ground ore. Initial scouting tests were carried out with oven-dried samples of "leach feed", i.e. feed to the pachucas, from the Quirke Mine at Elliot Lake (Rio Algom Ltd.). The leach feed was 50% minus 200 mesh (74 µm) and contained typically 0.1% U and 326 pCi/g Ra.

Scouting Tests

In recognition that uranium is more leachable than radium, the first scouting tests were carried out first only for radium. In the standard test for each salt, a 1.0 mol/L solution was contacted for 1.0 h with the ground leach feed (ratio 2.5 mL solution per g of ore) at 20°C. The contacts were carried out with 100g batches of solids in a 600 mL beaker agitated by a magnetic stirring bar. The pH of the solution was noted but no attempt was made to control it by buffering, etc. The leachants were ranked in order of the extract concentration as shown in Table II below.

Table II

Leachant	^{226}Ra in the Extract (pCi/mL)	Approx. % Ra Removal
Li Cl, $BaCl_2$, $Na_2S_2O_3$, Na_2SO_3, Na_2CO_3, H_2O	0 - 10	0 - 8
NaCl, KCl, NH_4Cl, $MnCl_2$, $AlCl_3$, NaBr, CH_3COONa, KNO_3	10 - 20	8 - 15
CsCl, $CaCl_2$, $K_4Fe(CN)_6^*$ (*Conc 0.5M)	20 - 30	15 - 23
$FeCl_3$	67	50

It will be seen that sulphur-containing salts are all poor leachants; this is to be expected in view of their tendency to oxidise to form sulphates in which radium is known to be very insoluble. Barium chloride is also a poor leachant, probably due to co-precipitation of radium with barium sulphate in the presence of traces of sulphate from the air oxidation of sulphides in the leach feed.

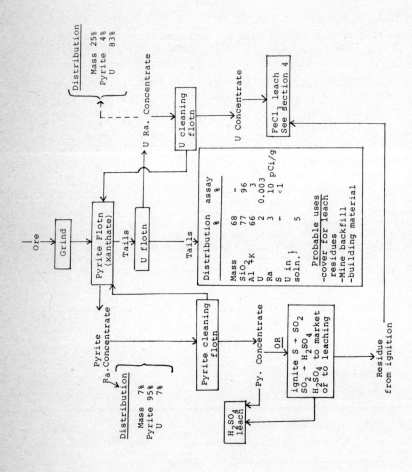

Figure 2 Conceptual flowsheet for uranium and pyrite flotation from Elliot Lake ores with a chelating agent, e.g. cupferron.

Ferric Chloride as Leachant

By far the best leachant is ferric chloride. Ferric ion is well-known as an effective leachant for many minerals [20]. In the case of uranium minerals, the ferric ion acts as an oxidant whereby quadrivalent uranium is converted to its more soluble hexavalent form as discussed in section 2 above. This mechanism is important in conventional sulphuric acid leaching; leaching of ores by ferric sulphate itself has been studied by many previous workers [see, e.g. 20,21].

Only one extensive previous investigation of uranium leaching by ferric chloride seems to have been published, namely the 1959 patent of Sawyer and Handley [22]. They employed approximately 0.4 mol/L leach solutions at $85^{\circ}C$ in contact with ores from the Colorado plateau. In one example with a 3.4% U ore, the leach filtrate was found to contain 97% and 89% of the uranium and radium values respectively.

The scouting tests (Table II) were followed by a more detailed series of $FeCl_3$ leaching tests in which the percentage extractions of both radium and uranium were determined over ranges of concentration (0.1-1.0 mol/L), temperature (47 - $74^{\circ}C$) and contact time (0.25 - 24 h). While the percentage uranium extraction increased with $FeCl_3$ concentration, temperature and time to a maximum value of 97%, the radium extraction was greatest at the lowest temperature ($47^{\circ}C$) and at contact times of about 1 hour. The highest radium extraction observed was only 67%.

The decrease of radium extraction at high temperatures and long contact times is believed to be due to the gradual formation of sulphate ions from the action of Fe^{3+} on pyrite, in addition to slow leaching of metal ions (Ba^{2+}, Pb^{2+}) which subsequently precipitate as sulphates, taking up radium as a coprecipitate. It should be noted also that some pyrrhotite (FeS) is present in the ore; this is oxidised quite rapidly by $FeCl_3$ giving rise to sulphate ion even in the early stages of the contact. These sulphides are not present to such a major extent in the Colorado ores, explaining the relatively good radium leaching results of Sawyer and Handley [22]. The redox processes involving pyrrhotite and pyrite were studied by EMF measurements. An increase in EMF and in uranium extraction was seen when sodium chlorate (0.6 g per 100 g feed) was added initially.

Flotation Pretreatment

The above results suggest that the leaching effectiveness of ferric chloride, particularly for radium, could be increased by removing sources of sulphate ion, i.e. pyrite and pyrrhotite. This can readily be done by flotation of the ground ore using a xanthate collector (see Section 3 above). Figure 3 shows a conceptual flowsheet based upon bench-scale experiments that have been carried out on the major processing steps indicated. It must be emphasised that as in the case of Figure 2, work is still in progress on several aspects of this flowsheet, and it presently provides a framework for our experiments rather than a firm recommendation for process development.

The sulphide flotation removes about 7% of the ore with essentially all sulphides and about 10% of the uranium. It is expected that a separate $FeCl_3$ leach in the presence of an oxidant can remove at least 90% of the uranium from the sulphide concentrate, but that radium removal will be very poor; therefore the relatively small amount of pyrite-rich leached solids should be stored in an inaccessible site.

However, the main process stream of interest is the "tail" from the sulphide flotation, consisting of 93% of the ore and about 90% of the uranium. Tests with this material at $74^{\circ}C$ and a liquid/solid ratio of 2.5 mL/g showed good radium removal at long contact times. For example, a single 24-hour contact with 0.10 mol/L ferric chloride solution reduced the uranium in the solids to 24 p.p.m. and the radium to 24 pCi/g. The residual radium levels in proportion to uranium in the leached solids were consistently greater than those in the ore and the sulphide float tails, indicating that radium is still relatively harder to leach, but the final radium concentration approaches the 20 pCi/g level that is considered [23] acceptable.

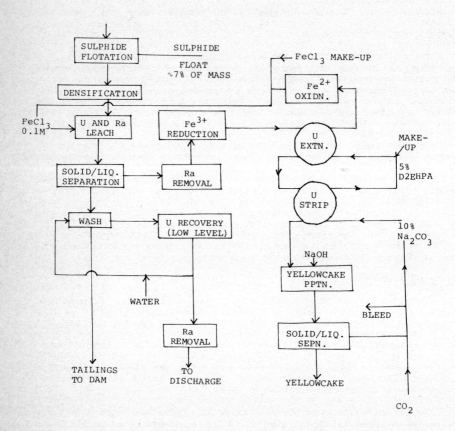

Figure 3   Conceptual flowsheet for uranium and radium leaching by means
of ferric chloride, with uranium recovery and ferric chloride
recycle.

Radium Removal from Leach Liquor

The filtered leach liquor contains uranium (~0.4 g/L), ferric chloride (~16 g/L) plus undetermined small amounts of rare earths, heavy metals, etc. The pH is approximately 1.0. It has been claimed [24] that manganese dioxide is a very effective adsorbent for radium, and a scouting test with the leach liquor has given 94% removal of radium. Further development is needed for the radium removal step (see also Section 5). Barium sulphate precipitation is not suitable because sulphate cannot be introduced into the circuit.

Uranium Removal from Leach Liquor

Uranium can be recovered effectively from the leach liquor by a modified DAPEX process [25], i.e. solvent extraction with a solution of di-2-ethylhexyl-phosphoric acid (5%) and tributyl phosphate (2.5%) in kerosene. The extraction must be preceded by a reduction process to convert Fe^{3+} to Fe^{2+} which is much less strongly extracted by the organic solvent than Fe^{3+}. This reduction was carried out in our tests with sulphur dioxide, but in a process it would be necessary to use iron or an electroreduction cell; sulphur dioxide addition would lead to an undesirable build-up of sulphate ions.

The percentage removal of uranium from a reduced leach liquor by solvent extraction was found to be in excess of 99% at an organic/aqueous phase ratio of 0.125. Stripping with 10% sodium carbonate solution to form uranyl carbonate in the aqueous phase is also more than 99% effective, at an O/A phase ratio of 4. An increase of the O/A ratio to 8 gave a sharply reduced effectiveness which may have been due to the depletion of the sodium carbonate.

The carbonate strip liquor did not contain a solid phase, indicating that large amounts of iron had not passed through the circuit.

Moving back to the ferric chloride circuit in Figure 3, the raffinate from the uranium extraction must be reoxidised to the ferric state; make-up ferric chloride is added and the stream is returned to the leaching step. The ferric chloride circuit is "bled" by the washing of the leached solids from which the main leach liquor stream has been separated. The wash liquor will contain low uranium values (<0.1 g U/L) which must be at least partially recovered by ion exchange or other suitable means. Radium may be removed from this liquor by barium sulphate coprecipitation prior to discharge; sulphate ions are of course no problem at this point.

It must be re-emphasised that Figure 3 is only a conceptual flowsheet. Although the main steps (sulphide flotation, leaching, ferric ion reduction, extraction and stripping) have been performed at a laboratory scale, further laboratory and engineering assessments are needed. An important question is; to what extent can the ferric chloride circuit be closed? The answer will determine not only the cost of ferric chloride make-up, but also the amount of chloride ion that must be discharged. Environmental regulations in Ontario set the chloride maximum level in effluents at 750 p.p.m. Rough hand-calculations indicate that the overall chloride levels in the liquid effluent will fall below this figure, but the flow sheet calculations must be developed using computer assistance, to explore the effect of process variables in greater detail.

5. Chelating Leachants for Sulphuric Acid Leach Discharge
(A. Corsini, D. Keller, A. Nixon and A. Pidruczny)

As noted in Section 2, the milling of uranium at Elliot Lake is based on the conventional sulphuric acid leach process. Skeaff [10] has reported that rapid dissolution and reprecipitation of radium-226 takes place in the first pachuca tank. The reprecipitated radium-226 is preferentially associated with the minus 200 mesh fraction of the solids and only about 0.2% of radium-226 remains in solution.

EDTA has been shown to be effective in the removal of Ra from U leach solids [6,23,26] but only with the use of levels of EDTA which would be prohibitively expensive in practice.

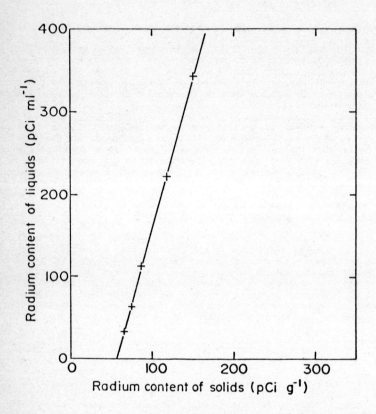

Figure 4   Radium distribution isotherm from tests at variable liquid to
            solid ratio.
            (L/S = 0.75 to 8, pH = 11, contact time = 1 h, [EDTA] = 0.04
            mol/L, temperature = 22°C).

Scope

There are three principal objectives of this part of the present study.

i) to determine the minimum concentration of radium in tailings achievable by extraction with EDTA, and if possible to meet the target level of 20 pCi/g.

ii) to optimize the extraction of radium with respect to the quantity of EDTA required.

iii) to determine the causes of consumption of EDTA for the purpose of minimizing the losses of EDTA and to develop a method for the recovery of EDTA.

Consideration has been given to other complexing agents, however none appeared to offer a better compromise between strong radium complexing ability and moderate cost. Hence, only EDTA was selected for study. A bulk price of $2.36/kg has recently been quoted for EDTA.

5.1 Leaching

The feed solids for the EDTA leach experiments (obtained from Quirke Mill, Rio Algom Mines Ltd.) were: fine tailings from countercurrent decantation (CCD, see Fig. 1), and final pachuca (#24) output. The solids were pre-washed with deionised water.

Both dried solid samples and liquid samples were submitted for analysis in standard jars which were sealed to prevent escape of radon gas and stored for 30 days to establish secular equilibrium. A series of standards was prepapred. The 609.4 keV photopeak of ^{214}Bi was monitored. The ratio of ^{214}Bi activities in the standard and each sample is identified to the ratio of ^{226}Ra activities.

Uncomplexed EDTA was determined by titration with Zn(II) in an ammonia buffer. The sum of uncomplexed EDTA and EDTA bound to alkaline earths was determined by titration with Zn(II) in an acetate buffer. The concentration of iron was determined by a spectrophotometric method.

Results and Discussion

Radium is extracted only when free EDTA is present. This is not unexpected as the formation constant of the radium-EDTA complex is less than that of almost every other metallic ion except those of the alkali metals [29].

The CCD fine tailings result from partial neutralization of the final pachuca output with lime or limestone as noted in Section 2 (Fig. 1). The high calcium content of these solids results in large EDTA consumption. Hence it was considered more advantageous to treat the final pachuca discharge before the partial neutralization. The radium content of the pachuca discharge solids used in this work was 326 ± 7 pCi/g.

A broad maximum in the radium extraction (~80%) was obtained around pH 10.

There was no significant difference observed between the extraction obtained in a few minutes and that obtained in 25 hours. However, an increase in the total consumption of EDTA with contact time was observed. This is accounted for primarily by the increased concentration of iron which probably occurs as a result of the oxidation of pyrite. Consequently, long contact times for the EDTA leach are disadvantageous.

At sulphate levels up to 0.35 mol/L and phosphate levels up to 0.071 mol/L no significant reduction in the extraction of radium by EDTA was observed.

A radium distribution isotherm obtained from leaches at variable liquid to solid ratios is illustrated in Figure 4. The relationship between the radium contents of the solids and of the liquids can be expressed as

$$x = x_0 + my$$

where x is the radium content of the solids, y is the radium content of the liquids and x_0 is an " unextractable" concentration of radium. The behaviour of the extractable fraction of radium is characteristic of a Langmuir adsorption isotherm at low concentration. An apparently "unextractable" fraction of radium has been reported previously by others [6,27,28].

Pachuca discharge previously leached with EDTA was further leached with sodium chloride, nitric acid, and hydrochloric acid to determine if the radium content could be reduced below the value of x_o. Aqueous NaCl was ineffective. However, hot, concentrated (70°C, 3M.) nitric or hydrochloric acid reduced the radium content of the solids significantly.

These results lend some support to the hypothesis that the "unextractable" fraction of the radium is trapped in a refractory material, one which is inert to EDTA or NaCl but which may be attacked by strong acid.

The consumption of EDTA is between 7 and 9 mmol per kg of solids. Of this, about 3 mmol/kg may be accounted for by complexation with iron. A similar quantity is accounted for by calcium and magnesium and 1 mmol/kg is consumed by lead.

No other metal ions are found in other than trace quantities. Therefore, about 2 mmols of EDTA/kg appear to be lost, probably by oxidation and by adsorption on the solids.

Conclusions

The major conclusions in regard to EDTA leaching are:

i) EDTA may be used to efficiently extract radium from Elliot Lake uranium leach discharge to approach the values of x_o using dilute solutions, relatively low liquid to solid ratio, ambient temperature and short contact times.

ii) The EDTA concentrations used are considerably lower than previously reported and well below the effective concentrations of inorganic salts required. This is because, in the present work, the solids are taken directly from the uranium leach, are washed thoroughly and thus have relatively low contents of soluble metals (e.g. Fe, Ca) which complex with EDTA.

iii) The liquid to solid ratios are much lower than previously reported for EDTA leaches and are also much lower than those required for salt leaching.

iv) Radium extraction by EDTA appears to be determined by an adsorption mechanism and is not adversely affected by the presence of sulphate. The use of EDTA would therefore avoid the sulphate buildup problem anticipated in a salt washing cycle [27].

v) The residual radium content of the tailings from EDTA leaching does not meet the target value of 20 pCi/g. However, this residual radium can only be removed by very vigorous conditions and it seems likely that its release into the environment would be correspondingly slow. This "unextractable" portion may be related to the mineralization of the ore and lower values may be obtained for uranium leach discharges from different sources.

5.2 Regeneration of Leachant Solutions

Compared to previous estimates, a considerable reduction in the quantity of EDTA required to extract radium efficiently from tailings has been achieved. The amount concerned, about 40 to 80 mmol/kg of solids (12 to 24 g of H_4EDTA/kg) is still too high to be used economically on a once-through basis. There are two distinct aspects to the recycling of EDTA from leachate solutions: radium must be recovered in a concentrated form for disposal as a medium or high level waste, and EDTA must be recovered for the purpose of recycling.

Experimental

Radium distribution coefficients were measured by contacting BioRad AG50W-X8, 100-200 mesh resin in the sodium form with leachate solution. The pH of the leachate solution was adjusted with aqueous sodium hydroxide or sulphuric acid as necessary to cover the pH range 12-4 in increments of approximately one pH unit. Radium measurement was carried out as described above. The distribution coefficient D', is defined here as the ratio of pCi of radium per gram of dry resin to pCi of radium per mL of solution.

Results and Discussion

Radium Recovery

The usual technique for removal of radium from liquids is by co-precipitation with barium sulphate. This method is used to treat the effluent from tailings ponds in Elliot Lake [30].

The use of barium sulphate precipitation has several serious drawbacks [27]. The kinetics of precipitation are slow and residence times of greater than one hour are required for maximum radium removal. Settling of the barium sulphate precipitate is also slow. Therefore, our approach has concentrated on the use of ion-exchange.

It is possible to derive, from the law of mass action, an expression which will describe the distribution coefficient of radium between a strong-acid (sulphonate) type resin, and an aqueous EDTA solution as a function of pH. The general relationship is given by

$$\log D' = n \log C - n \log \frac{[H^+]}{K_{H^+}^{M^{n+}/n}} + \frac{[Na^+]}{K_{Na^+}^{M^{n+}/n}} - \log \alpha_{M(Y)}$$

Here, D' is a distribution coefficient at infinitesimal loading of the resin, n is the charge of the metal ion, C is the capacity of the resin in milliequivalents per gram of dry resin,

$$K_{H^+}^{M^{n+}/n} \quad \text{and} \quad K_{Na^+}^{M^{n+}/n}$$

represent the equilibrium constants for the exchange of a metal ion M^{n+} with hydrogen ions and sodium ions respectively, and $\alpha_{M(Y)}$ is a complex function which quantifies the degree to which the metal ion is sequestered by EDTA. It is not possible to measure values of log D' very different from zero as in such cases either the radium content of the resin or the radium content of the aqueous phase is so low that statistical counting errors are excessive. Calculated and experimental values of log D' at pH 7,8 and 9 are compared in Table III.

Table III

Calculated and experimental value of log D' as a function of pH.

pH	log D' calculated	log D' experimental
7	1.42	1.37
8	0.40	0.45
9	− 0.60	− 0.24

The agreement between the calculated and experimental values is quite reasonable in view of the fact that the above equation does not take into account activities or variations in ionic strength. By careful control of pH, it should be possible to take up radium selectively and reject all other metal ions with the exception of barium.

Radium stripping from the resin may be accomplished with concentrated NaCl or possibly with EDTA solution at about pH 11 from which a high level waste could be prepared by precipitation with barium sulphate.

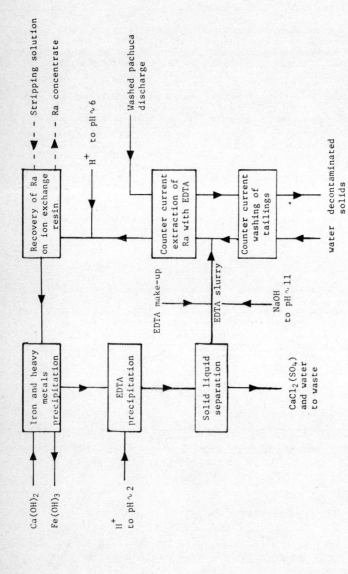

Figure 5   Conceptual flowsheet for radium extraction from uranium mill
            leach solids and EDTA recycle.

Comparison is needed between the ion exchange method as described above, and the biosorption techniques which are under investigation (Section 5).

EDTA Recovery

Since about 10% of the EDTA is used up by complexation it is evident that even if radium is removed, it will not be possible to recycle the EDTA solution very many times. Therefore, the recovery of complexed EDTA is imperative. Unfortunately, the differences in the chemistry of the alkaline earths and iron and lead make it difficult to recover EDTA from all complexed forms in a single-stage process. A two stage process involving lime treatment and acid precipitation is described below.

In the first stage, a quantity of lime slightly greater than the stoichiometric quantity of EDTA is added to the leachate. This addition of lime has two effects: It raises the pH of the leachate solution to slightly over 12 and the excess calcium displaces iron from its EDTA complex as ferric hydroxide. The precipitate obtained from lime treatment settles more rapidly than that obtained from sodium hydroxide treatment. Other heavy metals are precipitated in the same way. The supernatant liquid remaining after hydroxide precipitation is clear and almost colourless. It consists of an aqueous solution of calcium-EDTA with a small percentage of magnesium-EDTA at about pH 12. EDTA can be recovered from aqueous Ca-EDTA by acidification and precipitation. The minimum solubility of EDTA in sulphuric acid (0.13 g/L) is observed at about pH 1.8. EDTA cannot be precipitated by sulphuric acid from a solution initially containing greater than 0.014M Ca-EDTA without also precipitating calcium sulphate.

Flowsheet

From the elements described above, it is possible to construct the outline of a process to recycle EDTA. A conceptual flowsheet is shown in Figure 5. As in the case of the flowsheets shown in Figures 2 and 3, this conceptual flowsheet is tentative and does not constitute a firm recommendation for process development. Assuming a L/S ratio of 2 mL/g in the leaching step and an EDTA concentration of 0.04 mol/L, estimates of lime for the precipitation of iron, and sulphuric or hydrochloric acids for the precipitation of EDTA can be made on a stoichiometric basis.

Table IV

Quantities of Chemicals Required per Tonne of Solids

EDTA	29.8 kg* (as disodium EDTA dihydrate)
CaO	4.5 kg
H_2SO_4	15.7 kg**
NaOH	11.2 kg
Water	2,000 L

*   No recycle assumed

** 11.7 kg of HCℓ could also be used; however, the cost of HCℓ is considerably higher than that of H_2SO_4

The efficiency of EDTA recycling is dependent on the losses of EDTA from three sources: non-recoverable chemical loss (0.75 kg per tonne), soluble loss from the precipitation of EDTA (0.25 kg per tonne), and the loss of EDTA with leached solids.

Overall, it does not seem unreasonable to suppose that more than 90% (based on the conditions above) of the EDTA could be recycled, so that EDTA requirements could be in the order of 2 kg per tonne of solids treated.

6. Biosorptive Recovery of Radium (M. Tsezos, D. Keller)

Information available in the literature [31-33] has clearly indicated that microbial cells possess the ability to bind with certain cations and remove them from solution. This property is expressed equally well by living and dead microbial cells. The phenomenon of retention of cations from solution by dead microbial cells has been termed biosorption.

Work done by Tsezos during the last 5 years [34] has clearly indicated that inactive waste microbial biomass, byproduct of industrial fermentations or waste-water treatment operations, can effectively sequester radionuclides like uranium or thorium from waste solutions. The study of the equilibrium and kinetics of the uranium and thorium biosorptive sequestering has resulted in a better understanding of the processes involved, and has allowed the formulation of a biosorption mechanism hypothesis [35].

On the basis of the experience accumulated on the sequestering of uranium and thorium it was decided to examine biosorption as a potential sequestering process for radium. A variety of waste inactive microbial biomass types and synthetic materials such as activated carbons or ion exchange resins have been tested for their radium uptake capacity. The following parameters affecting the uptake of dissolved Ra by microbial cells are being studied.

i)   Solution pH: values of 2,4,7 and 10.

ii) Initial solution Ra concentrations of 50, 500 and 1000 pCi mL^{-1}.

iii) Liquid to Solid ratio: 200 to 10,000 mL/g.

iv) Type of biomass.

Strict control is exerted over all experimental conditions so that the observed radium uptakes can be attributed solely to biosorption and not radium uptake by the walls of the contact and processing systems. Contact time was 24 hours and contact temperature was 22°C.

Results

Radium uptake by microbial biomass appears to be highly specific to biomass type as some types (e.g. Aspergillus niger) exhibit negligible radium uptake while others exhibit radium removals of >99%.
Radium uptake has been demonstrated to be a strong function of solution pH. Minimal uptake is observed at pH = 2. Uptake increases with solution pH up to the 7 to 10 pH range.

Even at the highest Ra concentration and the greatest liquid to solid ratio used, saturation of the biomass was not obtained.

Future Work on Biosorption

The work continues towards detailed evaluation of the radium equilibrium uptake capacity of the most efficient biomass types selected, optimum conditions for biosorption, the associated kinetics and the ability of the biomass to sequester Ra from process solutions containing other competing cations (e.g. Fe, Zn) or complexing agents (e.g. EDTA). Typical solutions of this type are leach solutions from FeCl$_3$ or EDTA leaching as described in Sections 4 and 5 above.

The ultimate objective of this work is the development of a novel technology, based on biosorption, that will be able to selectively sequester radium from leach solutions. Such a process can be incorporated in a radium removal circuit such as that shown in Fig. 3.

7. Conclusions and Further Work

In sections 3-6 above, several possible means of removing radium from tailings by altering the milling process have been discussed. Table V below summarises the possible relationships between these phases of the project and the main areas where more technical data are needed.

Table V

Summary of Conclusions

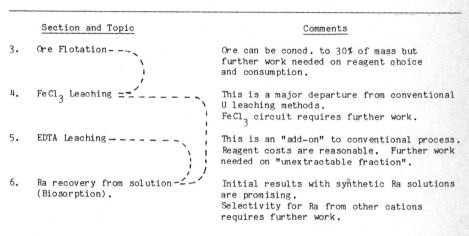

Section and Topic	Comments
3. Ore Flotation	Ore can be concd. to 30% of mass but further work needed on reagent choice and consumption.
4. $FeCl_3$ Leaching	This is a major departure from conventional U leaching methods. $FeCl_3$ circuit requires further work.
5. EDTA Leaching	This is an "add-on" to conventional process. Reagent costs are reasonable. Further work needed on "unextractable fraction".
6. Ra recovery from solution (Biosorption).	Initial results with synthetic Ra solutions are promising. Selectivity for Ra from other cations requires further work.

An additional comment, common to all the above sections, concerns the need for engineering design and economic analysis. It is necessary to have estimates of what the extra financial expenditure would be, in $ per tonne of ore, to produce environmentally acceptable solid tailings. Such estimates will call for flowsheet calculations and the design and costing of major equipment items, in comparison with "best cost" estimates for a conventional uranium mill. A first approach to this objective is currently being made in a design project involving the present authors, additional faculty colleagues and final-year chemical engineering students at McMaster University.

Finally, it must be emphasized again that all the present experiments have been made with Elliot Lake samples. Considerably different results could be expected with samples from other mills. It is hoped that the approaches developed in this work will be of interest for testing materials from other major uranium mining areas.

Acknowledgments

We are grateful to the Natural Science and Engineering Research Council of Canada for the financial support provided by a strategic grant, and to Rio Algom Ltd. for their cooperation in providing samples for test.

We are also grateful to K.D. Hester (Rio Algom Ltd.) and Dr. L.W. Shemilt for discussions and advice.

References

1. Runnals, O.J.C., Canada's Role as a Uranium Supplier, Opening Address, Course on Extractive Metallurgy of Uranium, University of Toronto (1978).

2. Bayda, E.D. (Chmn.), The Cluff Lake Board of Enquiry, Final Report, Regina, Sask. (May 31, 1978).

3. Caverly, D.S. (Chmn.), A Public Hearing by the Environmental Assessment Board into the Expansion of the Uranium Mines in the Elliot Lake Area, Final Report (May 1979).

4. Itzkovitch, I.J. and Ritcey, G.M., Removal of Radionuclides from Process Streams - A Review, CANMET Report 79-21, Energy Mines and Resources Canada, Ottawa (1979).

5. Phillips, C.R. and Poon, Y.C., Status and Future Possibilities for the Recovery of Uranium, Thorium and Rare Earths from Canadian Ores, with Emphasis on the Problem of Radium, Minerals Sci. Eng., 12, 53-72 (1980).

6. Yagnik, S.K., Baird, M.H.I. and Banerjee, S., An Investigation of Radium Extraction from Uranium Mill Tailings, Hydrometallurgy 7, 61-75 (1981).

7. Raicevic, D., Decontamination of Elliot Lake Uranium Tailings, CIM Bulletin 72 (808), 109-115 (1979).

8. Hester, K.D., Current Developments at Rio Algom Elliot Lake, CIM Bulletin, 72 (804), 181-188 (1979).

9. LaRocque, E. and Pakkala, E., Current Leaching and Product Recovery Practice at Denison Mines Ltd., CIM Bulletin, 72 (804), 172-176 (1979).

10. Skeaff, J.M., Survey of the Occurrence of Ra-226 in the Rio Algom Quirke I Uranium Mill, Elliot Lake, CIM Bulletin, 74 (830), 115-121 (1981).

11. Skeaff, J.M., Distribution of Radium-226 in Uranium Tailings, Report MRP/MSL 77-340(J), CANMET, Energy Mines and Resources Canada (1977).

12. Eigeles, M.A., Grekulova, L.A., Volava, M.L., Shishova, A.M. and Laksanko, V.M., Proceedings, 2nd UN Int. Conf. on Peaceful Uses of Atomic Energy, 3, pp. 147-153 (1958).

13. Korchinski, I.J.O., Craig, G.A., Cavers, S.D. and Van Cleave, A.B., Beneficiation of Low Grade Uranium Ores, Chem. in Canada 6, 34 (1954).

14. Crawford, L.W., Gunn, B., Cavers, S.D. and Van Cleave, A.B., Beneficiation of Low Grade Saskatchewan Uranium Ores IV, Can. J. Chem. Eng., 35, pp. 99-104, (Oct. 1957).

15. Van Cleave, A.B., Beneficiation of Low Grade Pegmatitic Uranium Ores, Can. Min. Metall. Trans. 59, pp. 433-440 (1956).

16. Estefan, S.F. and Malati, M.A., Activation of Oleate Flotation of Quartz by Alkaline Earth Ions. Trans. Inst. Min. & Metall., Sect. C. 82, pp. 225-228 (1973).

17. Dunn, Robert W., Flotation Preconcentration of Elliot Lake Uranium Ore, Rio Algom Report R74-03 (May 1974).

18. Shakir, K., Two-stage Flotation Process for the Concentration of the Uranium-bearing Ore from Gebel Katrany, Egypt, Ind. J. of Technology 12, pp. 304-306 (1974).

19. Marabini, A.M. and Rinnelli, G., Flotation of Pitchblende with a Chelating Agent and Fuel Oil, Trans. Inst. Min. and Metall., Section C, 82, pp. 225-228 (1973).

20. Dutrizac, J.E. and MacDonald, R.J.C., Ferric Ion as a Leaching Medium, Minerals Sci. Eng. 6, 59-100 (1974).

21. Ring, R.J., Ferric Sulphate Leaching of Some Australian Uranium Ores, Hydrometallurgy 6, 89-101 (1980).

22. Sawyer, C.W. and Handley, R.W., Process of Extracting Uranium and Radium from Ores, U.S. Patent 2, 894, 804, July 14 (1959).

23. Borrowman, S.R. and Brooks, P.T., Radium Removal from Uranium Ores and Mill Tailings, U.S. Bureau of Mines Rept. of Inv. 8099 (1975).

24. Bland, C.J., A New Use for Deep Sea Nodules, CIM Bulletin, 72 (812), 29 (1979).

25. Merritt, R.C., The Extractive Metallurgy of Uranium, p. 204-205, Colorado Sch. of Mines Res. Inst. (1971).

26. Seeley, F.G., Problems in the Separation of Radium from Uranium Ore Tailings, Hydrometallurgy, 2, pp. 249-263 (1976-1977).

27. Ryan, R.K. and Levins, D.M., Extraction of Radium from Uranium Tailings, CIM Bulletin, 73 (822), pp. 126-133 (1980).

28. Levins, D.M., Ryan, R.K. and Strong, K.P., Leaching of Radium from Uranium Tailings. Proceedings of OECD/NEA Seminar on Management, Stabilization and Environmental Impact of Uranium Mill Tailings, Albuquerque, N.M. (July 1978).

29. Ringbom, A., Complexation in Analytical Chemistry, Interscience Publishers, New York/London (1963).

30. Moffett, D., The Disposal of Solid Wastes and Liquid Effluents from the Milling of Uranium Ores, CANMET Report 76-19 (1976).

31. Polikarpov, G.G., Radioecology of Aquatic Organisms, North Holland Publishing Co., Reinhold Book Division, New York (1966).

32. Chiu, Y.S., Recovery of Heavy Metals by Microbes, Ph.D. Thesis, University of Western Ontario, London, Ontario (1972).

33. Beveridge, T.J., The Response of the Cell Walls of Bacillus Subtilis to Electron Microscopic Stains, Can. J. Microbiol., 24, pp. 89-1104 (1978).

34. Tsezos, M. and Volesky, B., Biosorption of Uranium and Thorium, Biotechnology Bioengineering, 23, pp. 583-604 (1981).

35. Tsezos, M. and Volesky, B., Biosorptive Concentration of Nuclear Fuel Elements, Presented at the Annual A.C.S. Meeting, Las Vegas, NV (August 1980).

A METHOD OF COMPUTING THE RATE OF LEACHING OF RADIO-NUCLIDES

FROM ABANDONED URANIUM MINE TAILINGS

C.J. Bland
The University of Calgary
Calgary, Alberta, Canada   T2N 1N4

ABSTRACT

In order to determine how fast radionuclides leach out of mine tailings, a simple mathematical model has been constructed. A set of coupled differential equations include the effects of loss by leaching, radioactive decay, and growth from precursors. Using a desk top microcomputer, leaching rates can be derived if data are available providing the radionuclide concentrations for fresh tailings and similar tailings which have undergone leaching for a known time since they were deposited. When further data become available such effects as seasonal variations in leaching could be easily incorporated in the model.

Introduction

        After the extraction of the U_3O_8 (yellowcake) from ore, most of the
mill feed is deposited as tailing. The uranium industry in Canada generates
several thousand tons of these tailings per day. [1]. The radioactive
daughters of uranium such as the long lived radionuclides ^{230}Th, ^{226}Ra, ^{210}Pb are
not removed by current milling procedures and are therefore discharged along with
the tailings. The growing concern regarding the proper disposal of radioactive
tailings has been the subject of much discussion. [2]. One of the most
 important questions to be answered is at what rate are uranium daughter nuclides
removed by leaching. In order to provide some approximate estimates, we have
collected data on fresh and twenty year old tailings from Elliot Lake, Ontario
and used this data to compute the rate of leaching of long lived uranium
daughters.

Radionuclide Concentrations in Tailings

        Activities of ^{230}Th, ^{226}Ra, ^{210}Pb and ^{210}Po as well as other short
lived radionuclides are found in tailings as unreacted silicous gangue together
with a small amount of fine grain gypsum produced from the lime used for neutra-
lisation of the slurry. Little information is available on the radioactivity of
tailings of well established age. However, Moffett and Tellier [3] have
provided data on twenty year old tailings. These tailings from an abandoned mine
at Elliot Lake cover an area of 16 hectares. The tailings contain up to 4.5%
of pyrite, which, in the absence of residual basicity, oxidizes to form sulphuric
acid. This area is being studied by the Canada Centre for Mineral and Energy
Technology (CANMET) at Elliot Lake in order to develop the optimum measures
required to minimize run-off and to reclaim the tailings for the establishment of
vegetation. The reader is referred to Moffett and Tellier's paper for details of
the hydrology of the area as well as the composition of acidic run-off from
various depths within the tailings. In Table 1, their data for radionuclide
concentrations are reproduced. It should be noted that only estimates are available for the radionuclide concentrations in the composite tailings when they were
fresh, whereas the data for the older tailings are divided into slime and sand
fractions.

TABLE 1

Sample	Concentration pCi/g			
	^{230}Th	^{226}Ra	^{210}Pb	^{210}Po
Slimes (65% passing 75μm)	12.9	314	130	77.4
Sands (35% passing 75μm)	8.3	166	36	17.8
Estimate Fresh Composite	344	344	344	344

In order to find out whether the slimes and sands in modern tailings differ in
radionuclide concentrations, we collected samples of fresh tailings at a nearby
operating mine. The results of analysis are given in Table 2.

TABLE 2

Sample	Concentration pCi/g			
	^{230}Th	^{226}Ra	^{210}Pb	^{210}Po
Fresh Tailings (>75μm)	469±40	410±10	425±50	430±40
Fresh Tailings (<75μm)	402±40	456±10	435±50	445±40

Analysis performed by Monenco Analytical Laboratories. Errors estimated on basis
of counting statistics and duplicate measurements.

Apart from a slightly higher ^{226}Ra content in the finer tailings, there is no significant fractionation between fresh sands and slimes. The higher concentrations of activities recorded in modern tailings compared with the value given in Table 1 is presumably a reflection of the higher grade of ore being milled at the time the recent samples were taken.

Method for Calculating the Rate of Leaching

For a given radionuclide belonging to the uranium decay chain, the activity concentration will vary due to three major processes.

a) Growth of activity due to the decay of the 'mother' radionuclide.

b) Radioactive decay.

c) Removal by processes such as erosion and leaching.

Processes (a) and (b) may be simply expressed as the standard differential equations of decay and ingrowth. Process (c) may be taken into account by an extra term representing loss by leaching of a constant fraction of nuclei per unit time. (An example of a similar expression applied to non-radioactive leaching is given by Chapelle [4]). Thus for a given radionuclide concentration we have an equation:

$$\frac{dN_j}{dt} = -\lambda_j N_j + \lambda_{j-1} N_{j-1} - K_j N_j$$

Where N_j, N_{j-1} are the concentrations of a given nuclide and its precursor in units such as nuclei/Kg.

λ_j, λ_{j-1} are the radioactive decay constants (i.e. ln2/half-life) of a given nuclide and its precursor.

K_j is the rate of loss of nuclei by leaching, etc.

If we restrict our attention to only those uranium daughter nuclides with half lives exceeding several weeks, i.e. ^{230}Th, ^{226}Ra, ^{210}Pb, and ^{210}Po, then we will have a set of four simultaneous differential equations. Now the first equation referring to the activity of ^{230}Th contains only two terms on the left hand side, i.e.

$$\frac{dN_1}{dt} = -\lambda_1 N_1 - K_1 N_1$$

^{230}Th can be considered unsupported because the parent ^{234}U was presumably removed in the mill. Given the concentration of ^{230}Th activity estimated for fresh tailings and for ~20 year old tailings, the above equation may be integrated to give the rate of leaching, i.e.

$$K_1 = \frac{\ln\{N_1(0)/N(t)\}}{t} - \lambda_1$$

where $N_1(0)$ and $N_1(t)$ are the concentrations of ^{230}Th nuclei in fresh and old (age = t) tailings respectively.

Knowing K_1, the leaching rate of ^{230}Th, we may proceed to find the corresponding parameter, K_2, for ^{226}Ra. This has been done by numerical integration on a small desk top microcomputer (PET$^{(TM)}$ model 2001). The incremental change in the concentration of ^{226}Ra nuclei is given by:

$$\Delta N_2 = (\lambda_1 N_1 - \lambda_2 N_2 - K_2 N_2) \Delta t$$

Since the concentration of ^{230}Th varies with time as

$$N_1(t) = N_1(0) e^{-(\lambda_1 + K_1)t}$$

then the variation of the concentration of ^{226}Ra nuclei with time may be found by

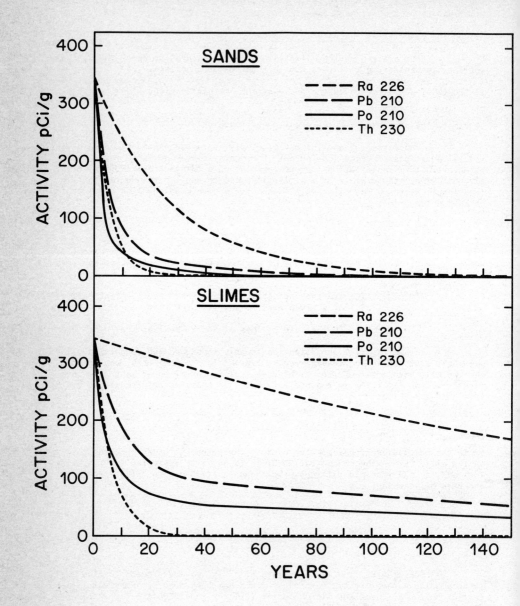

numerical integration for various values of the ^{226}Ra leaching rate K_2. This constant is adjusted until the predicted value of the ^{226}Ra nuclei concentration in old tailings tends to the observed value. Once K_2 is known, the procedure is repeated for ^{210}Pb and ^{210}Po, to yield the optimum estimates of K_3 and K_4.

Results of the Calculation

The concentrations of activities in fresh sands and slimes were assumed to be the same as the values given by Moffett and Tellier for composite tailings, i.e. 344 pCi/g. The calculations were done separately for slimes and sands to yield two sets of values for the leaching of half lives which are given in Table 3.

TABLE 3

Leaching Half Lives

Sample	^{230}Th	^{226}Ra	^{210}Pb	^{210}Po
Slimes	4.2y	162.7y	9.6y	0.54y
Sands	3.7y	19.1y	4.0y	0.37y

The leaching half life analogous to the radioactive half life is defined in terms of the leaching constants K_j. The latter represent the fraction of the radioactive nuclei or activity removed per unit time. The leaching half lives are given by:

$$\tau_j = \frac{\ln 2}{K_j}$$

Thus the activity of an unsupported radionuclide will decay with a total half life given by:

$$\frac{1}{\tau_{total}} = \frac{1}{\tau_{decay}} + \frac{1}{\tau_{leaching}}$$

Thus in the case of ^{230}Th and ^{226}Ra with radioactive half lives of 80,000 and 1,600 years respectively, loss by leaching is the dominant process whereas in the case of ^{210}Pb and ^{210}Po both decay and leaching are competitive processes.

Discussion of Results

One of the most interesting results obtained is the fact that ^{230}Th is leached rather rapidly from the tailings. As has been pointed out, however, by Langmuir and Herman [5], thorium is released from solids in acidic sulphate environments and remains mobile in the ground waters chiefly as sulphate complexes.

The calculations reveal that radium has the longest half life against leaching and is more effectively retained in the slimes than in the sands. This would indicate that radium is absorbed on the surface of the small grains and is held rather effectively in the present oxidising conditions.

It is of interest to predict the radionuclide composition of these tailings in the future. Figure 1 shows the predicted behaviour over a century from the time of deposition (assuming that conditions existing over the last 20 years persist). Due to the simple assumptions made which were dictated by lack of regular measurements on tailings at frequent intervals, all loss mechanisms, e.g. leaching, erosion, radon emanation, etc., are represented by a single term describing loss of nuclei at a constant rate. A more realistic calculation would take into account the variation of loss mechanisms with time due to seasonal change, variations in acidity, etc. It would also be useful to monitor long lived nuclei from other decay chains such as ^{231}Pa, ^{227}Ac in the ^{235}U series since run-off from old tailings contains substantial $^{223}Ra/^{226}Ra$ activity ratios.

Conclusions

The calculations show that useful information on the mobility of radionuclides in tailings may be obtained by regular monitoring of their concentrations of various radionuclides. Such monitoring should be considered an essential feature of the responsible management of such tailings.

Acknowledgements

This work was supported by a research agreement from CANMET, Energy, Mines and Resources, Canada. I wish to thank the staff of Rio Algom Ltd., Elliot Lake, for its collaboration, and in particular, Dr. Duncan Moffett for his valuable advice and assistance.

References

[1] Levins, D.M.: "Environmental Impact of Uranium Mining and Milling in Australia", CIM Bulletin, October 1980, 119-125 and references therein.

[2] Moffett, D.: "Uranium Waste Researchers Consider Alternate Means of Tailing Disposal", CIM Bulletin, January 1977, 48-50.

[3] Moffett, D. and Tellier, M.: "Radiological Investigations of an Abandoned Uranium Tailings Area", J. Environ. Qual. 7, 310-314 (1978).

[4] Chapelle, F.H.: "A Proposed Model for Predicting Trace Metal Composition of Fly-Ash Leachates", Env. Geol. 3, 117-122 (1980).

[5] Langmuir, D. and Herman, J.S.: "The Mobility of Thorium in Natural Waters at Low Temperatures", Geochim. Cosmochim. Acta 44, 1753-1766 (1980).

PRELIMINARY EVALUATION OF URANIUM MILL TAILINGS CONDITIONING AS AN ALTERNATIVE REMEDIAL ACTION TECHNOLOGY

D. R. Dreesen[1], E. J. Cokal[1], E. F. Thode[2],
L. E. Wangen[1], and J. M. Williams[1]

[1] Environmental Science Group, MS-495
Los Alamos National Laboratory
Los Alamos, New Mexico 87545

[2] Department of Management
New Mexico State University
Las Cruces, New Mexico 88003

ABSTRACT

Conditioning of uranium mill tailings is being investigated as an alternative remedial action for inactive tailings piles to be stabilized by the U.S. Department of Energy. Tailings from high priority sites have been characterized for elemental composition, mineralogy, aqueous leachable contaminants, and radon emanation power to provide a baseline to determine the environmental hazard control produced by conditioning. Thermal stabilization of tailings at high temperatures and removal of contaminants by sulfuric acid leaching are being investigated for technical merit as well as economic and engineering feasibility.

Table I

Elemental Composition of Uranium Mill
Tailings Compared With New Mexico Soils

Element or Isotope	Minimum Tailings Conc.[a]	Maximum Tailings Conc.[a]	Conc. Ratio Max. Tailings/Max. Soil
230Th	60 pCi/g	2190 pCi/g	810[b]
226Ra	316 pCi/g	2020 pCi/g	780
210Pb	295 pCi/g	2430 pCi/g	
Mo	7	800	420
Ag	< 0.5	12	240[b]
Cr	20	7880	170
Se	< 8	200	170
Cl	< 30	6730	130
U	33	470	120
V	150	4130	64
Cd	0.1	21	60
Sb	1.4	45	56
As	30	320	46
W	< 7	69	46[b]
Ba	360	21800	45
Cu	20	850	36
Zn	46	1780	35
Ga	< 22	370	19[b]
Co	1.2	110	10
Na	0.14%	3.7%	9
Fe	0.29%	13.1%	4
Ce	23	290	4
Ca	1.3%	7.7%	4
Mn	31	2040	3
K	< 0.5%	2.7%	1
Rb	< 23	130	1
Ti	830	3330	1
Mg	< 0.1%	1.1%	1
Al	1.3%	6.3%	1

[a]Concentration in μg/g unless otherwise indicated.
[b]Median soil concentration reported by Bowen [14].

INTRODUCTION

The goal of current uranium mill tailings management technologies is the long-term containment of mobile contaminants. The principal concern is the control of radon-222 releases and water leachable components in tailings. The perceived hazards resulting from such releases have prompted the U.S. Environmental Protection Agency to propose very stringent standards for the ultimate disposal of tailings at inactive mill sites [1]. The proposed standards would require a radon flux limit of 2 $pCi/m^2/s$ and prohibit the degradation of both surface and groundwater resulting from the leaching of radioclides and trace elements.

The most frequently proposed approach to the management of uranium mill tailings relies on barrier systems to contain hazardous contaminants and prevent their movement into the environment. The Los Alamos National Laboratory is investigating a different approach to tailings management as part of the technology development program of the US Department of Energy's Uranium Mill Tailings Remedial Action Project. The physical structure and/or chemical composition of tailings are being altered to either immobilize contaminants or remove contaminants before disposal. These physicochemical modifications are broadly termed "conditioning" methods.

CHARACTERIZATION

Individual tailings samples have been collected as near surface material (0-2m depths) from inactive tailings piles at Shiprock, New Mexico; Salt Lake City, Utah; Durango, Colorado; and Ambrosia Lake, New Mexico. These samples were analyzed for pH/Eh, neutralization equivalents, moisture content, water leachable constituents, and elemental composition by neutron activation analysis. The results of these analyses were used for producing composite samples from each tailings site.

The composites were subjected to intensive characterization before conditioning experiments were initiated. Neutron activation analysis, atomic absorption spectro-photometry, and gamma-ray spectroscopy were used to determine the composition of the tailings solids. These results are summarized in Table 1 and illustrate the obvious enrichment of ^{238}U and its daughters (i.e. ^{230}Th, ^{226}Ra, and ^{210}Pb). In addition, a wide variety of trace and minor elements are found to be highly enriched in tailings from these 4 sites including Mo, Ag, Cr, Se, V, Cd, and As. The results of the radionuclide analysis by gamma-ray spectroscopy have shown the depletion of ^{230}Th in tailings from sulfuric acid leach processes to levels 50% or less than that expected if in secular equilibrium with ^{238}U. The ^{210}Pb and ^{226}Ra measurements indicate no significant non-equilibrium in the lower part of ^{238}U decay series. (For further details see reference [2].)

The mineralogy of these composite tailings was determined by x-ray diffraction analysis and can be summarized as follows:
a) Shiprock tailings (acidic)
   - sands are primarily quartz (90%) with some gypsum (7%)
   - fines contain quartz (32%), gypsum (34%) and appreciable clays (12%) and plagioclase
b) Salt Lake City tailings (acidic)
   - sands are primarily quartz (68%) with some gypsum (14%) and minor amounts of clays, illite, and feldspars
   - fines contain quartz (48%), a little gypsum (7%), appreciable clays (11%) and illites (11%)
   - ferrophos vanadium residues contain some quartz (36%) but have substantial α-hematite and clino-choroapatite contents;
c) Durango tailings (alkaline)
   - sands are primarily quartz (88%) with some feldspars
   - fines contain some quartz (22-45%) with unidentified plagioclase-type minerals and non-aluminum silicates making up a majority of the remaining material;
d) Ambrosia Lake tailings (alkaline)
   - fines contain appreciable quartz (56%), clays (23%), and feldspars as well as calcite and kaolinite.

The broad classification of sands and fines was based on particle size analysis by wet sieving of disaggregated tailings. The sands contain sand-sized particles (50-80%)

Table II

Range of Water Leachable Components in Tailings Compared With Water Quality Standards or Criteria

	Minimum Conc. (mg/1)	Maximum Conc. (mg/1)	Water Quality Standard [h] (mg/1)
H⁺ (as pH)	9.4 f	2.7 d	–
SO_4^{2-}	15 f	3140 a	250
Ca	11 f	710 a	240
Cl	0.8 c	360 d	1300
Na	3 c	344 d	800
Al	< 0.1	180 d	5
Mg	6 f	98 d	90
NO_3^-	< 1	75 b	10
F	0.2 e	46 d	2
Fe	< 0.1	46 dd	0.3
V	0.3 a	24 e	2.5
Cu	< 0.1	22 d	1.0
K	3 g	13 b	–
Zn	< 0.1	11.8 d	5.0
U	< 0.05 e	8.6 c	0.03
Mn	< 0.1 f	7.3 d	0.05
PO_4^{3-}	< 2	7 e	–
Mo	< 0.1 f	5.4 d	0.05
Ni	< 0.01	1.6 d	0.2
Co	< 0.1	1.4 d	0.05
Pb	< 0.05	1.2 d	0.05
Cr	< 0.01 e	0.49	0.05
Be	< 0.01	0.40 c	0.03
Se	< 0.1	0.2 b	0.01
Cd	< 0.01	0.16 d	0.01
Sb	< 0.1	0.1	7.5
Ba	< 0.1	0.1	1.0
Ag	< 0.01	0.01	0.05

a - Shiprock sands
b - Shiprock fines
c - Salt Lake City sands
d - Salt Lake City fines
e - Salt Lake City ferrophos
f - Durango fines
g - Durango sands
h - See references [4, 5, 6, and 7]

with an appreciable content of silt-sized particles (20-50%). The fines are primarily silts (50-80%) with minor clay content (1-12%).

Electron microprobe analyses were used to examine the composition of individual tailings particles in an effort to infer mineralogy. Such studies on Shiprock fines indicated substantial quartz content (Si,O) with minor amounts of clays (K,Al,Si,O); gypsum (Ca,S,O); feldspars (Na,Ca,K,Fe,Al,Si,O); iron oxides (Fe,Mn,O); and, coalified humic substance (C,S). (More details on the mineralogy of these tailings can be found in reference [3].)

The emanating ^{226}Ra (the amount of ^{226}Ra which contributes ^{222}Rn to the gaseous surroundings) was measured for these tailings composites and ranges from 39-140 pCi/g for sands and 125-473 pCi/g for fines. For Shiprock and Salt Lake City acid tailings, the emanation coefficients (emanating ^{226}Ra/total ^{226}Ra) range from 10 to 12% except for the ferrophos residues which had a value of 20%. The Durango and Ambrosia Lake tailings from alkaline leach mills have emanation coefficients of 19 to 41%.

The leachable components in these tailings were determined by shaking with deionized water for 24 h (solid to liquid mass ratio of 1:5). The results of the analyses of these leachates are presented in Table II as minimum and maximum concentrations compared with water quality standards or criteria [4, 5, 6, 7].

Concentrations of all listed elements, except Ag, Ba, K, Mg, Na, and Sb, exceed these criteria values by sufficient amounts to warrant additional scrutiny. The following elements are high only in the leachates form acid tailings: Al, Be, Cd, Co, Fe, Ni, Pb, and Zn; only V is particularly high in alkaline tailings leachates. Uranium is highly water leachable from Salt Lake City uranium tailings and somewhat leachable from Durango Tailings. Molybdenum is present in high concentrations in leachates of two Salt Lake City tailings and Cr is high in Salt Lake City fines. For most elements (particularly acid mobile species), Salt Lake City fines gave the highest leachate concentrations of the tailings under study. This tailings composite generates the most acidic leachate (pH = 2.7) and has a fine grained texture (86% silt and clay).

The trends in elemental leachability reflect the qualitative pH behavior expected for these elements. Concentrations of most transition and heavy metals are controlled at low levels by insoluble hydroxides or carbonates. In contrast, the chemistry of some elements, such as Mo, Se, U, and V, suggests their presence as negatively charged ions or complexes that could remain in solution under alkaline conditions. Thus, the uranium extraction process, acid or carbonate, may be of paramount importance in determing mobile elements in tailings. Mineralogy of the uranium ores is undoubtedly an additional important factor controlling the aqueous mobility of these elements. The variability in elemental mobility at each site illustrates the complexity in predicting how "average" tailings might interact with surface or ground water. What commonality there is appears to be primarily a result of the inherent pH of the material and its particle size. The influence of particle size on mobility reflects surface area and the tendency for the fine-grained material to have higher contaminant concentrations.

THERMAL STABILIZATION

Our investigations of contaminant immobilization have centered on a process to radically modify the structure of tailings by treatment at high temperatures i.e., thermal stabilization. Sands and fines from the Shiprock and Durango tailings pile have been sintered at temperatures from 500° to 1200°C. The emanating ^{226}Ra for both Shiprock and Durango tailings as well as major mineral forms present in the sintered samples have been determined in order to relate emanating ^{226}Ra to changes in minerals.

The emanating power of these tailings as a function of sintering termperature is summarized in Table III. For Shiprock fines with emanating ^{226}Ra originally at 214 pCi/g, most of the reduction occurred between 700 and 1000°C (134 to 8.3 pCi/g); however, sufficient reduction (i.e., to 1-2 pCi/g) required temperatures greater than 1100°C for both Shiprock fines and sands. As seen in Table III, the Durango tailings showed much greater emanation reduction than the Shiprock samples, up to 800°C. As with the Shiprock samples, the Durango tailings required sintering temperatures of 1100°C or greater to reduce the emanation power to 1 - 2 pCi/g. The reduction in emanating ^{226}Ra at 1100° and 1200°C is greater for Durango tailings.

Table III

Percent Reduction in Emanating ^{226}Ra for
Sintering Temperatures from 500° to 1200°C

Sintering Temperature (°C)	Shiprock Sands[a] (%)	Shiprock Fines[a] (%)	Durango Sands[a] (%)	Durango Fines[a] (%)
500	15	16	48	61
600	29	27	64	68
700	44	37	76	80
800	63	58	87	88
900	83	84	92	91
1000	92	96.1	95.5	92
1100	96.4	98.8	99.0	99.8
1200	97.7	99.2	99.5	99.8

[a]Initial emanating ^{226}Ra: Shiprock sands   39 pCi/g
         Shiprock fines   214 pCi/g
         Durango sands   140 pCi/g
         Durango fines   473 pCi/g

Table IV

Extraction of Tailings Constituents From Durango Fines
as Influenced by Acid Concentration and Equilibrium
pH of Leaching Solution

Initial Sulfuric Acid Conc.	Final pH	Alkaline Earth Elements and Radionuclides (% Extracted)					
		Ba	Ca	Mg	^{226}Ra*	^{210}Pb*	^{230}Th*
Conc.	–	58	38	10	70	50	80
10N	–	1	2	34	10	0	60
1N	0.9	1	10	26	–	–	–
0.1N	6.5	1	10	10	–	–	–
0.01N	8.1	1	5	1	–	–	–
0.001N	9.1	1	1	1	–	–	–
O (H$_2$0)	9.4	1	1	1	–	–	–

Initial Sulfuric Acid Conc.	Final pH	Transition Metals Plus Al and U (% Extracted)								
		Fe	Al	Co	Cu	Mn	Ni	U	V	Zn
Conc.	–	3	1	–	69	21	9	52	37	–
10N	–	4	9	46	82	70	18	97	53	69
1N	0.9	1	3	42	82	32	14	65	46	67
0.01N	6.5	1	1	12	1	11	3	1	9	1
0.01N	8.1	1	1	1	1	1	1	11	1	1
0.001N	9.1	1	1	1	1	1	1	4	1	1
O (H$_2$0)	9.4	1	1	1	1	1	1	3	2	1

*Fines from small pile, other results on large pile fines

The weight loss due to heating ranged from 11% at 500°C to 24% at 1200°C for Shiprock fines. Corresponding losses for Shiprock sands were 2 to 5%. Durango fines ranged from 3 to 6% and Durango sands from 1 to 2%. It is assumed that most of this weight loss resulted from the loss of waters of hydration or decomposition of sulfates or carbonates.

The mineral transformations occurring over this range of temperatures were determined by x-ray diffractometry. The inferred changes in the mineralogy of Shiprock fines are as follows:

(1) gypsum disappears before 500°C (i.e., $CaSO_4 \cdot 2H_2O \rightarrow CaSO_4$ at 163°C); anhydrite is found at 500°C and above, but appears to start decreasing at 1000°C and is reduced by 75% at 1200°C; calcium silicates or aluminosilicates may be formed;

(2) quartz does not change substantially until 1200°C where about 60% is transformed perhaps to cristobalite (synthetic SiO_2) or to calcium silicates;

(3) clay minerals including illite disappear at 900°C;

(4) kaolinite seems to disappear by 500°C;

(5) new minerals (possibly plagioclase-type) are forming at about 900°C where the clay minerals and albite are disappearing;

(6) the amorphous content seems to substantially increase at 1100° to about twice the original content.

The gypsum and anhydrite response for Shiprock sands parallels that for fines. Quartz seems somewhat reduced (10%) only at 1200°C. Calcium silicates and cristobalite appear to be the high temperature mineral products.

The reduction in emanation power between 700 and 1000°C for Shiprock fines seems to correspond with the destruction of clays; however, to reach emanating ^{226}Ra levels of 1 pCi/g apparently requires significant melting as indicated by the production of increased amorphous character (i.e., glass). In addition, the apparent formation of calcium silicates and/or calcium aluminosilicates at temperatures of 900°C or greater could also immobilize radium in structures that limit the escape of radon gas.

The mineralogical changes in Durango fines resulting from sintering can be summarized as follows:

(1) quartz was reduced by 1/2 at 1200°C;

(2) gypsum disappears before 500°C; anhydrite forms, but begins to decrease between 1000° and 1100°C;

(3) barite is somewhat decreased at 1100°C and disappears by 1200°C;

(4) plagioclase-type minerals are reduced somewhat at 1100°C; but almost disappear by 1200°C;

(5) amorphous material has substantially increased at 1200°C (4-6 times original);

(6) obvious new mineral forms are not apparent.

As with the Shiprock tailings, fewer mineral transformations are apparent with the Durango sands. Some decrease in quartz and plagioclose occurs at 1200°C. Albite seems to disappear at 1200°C. There is some indication that cristobalite may be forming at 1100 to 1200°C. Amorphous character seems to increase somewhat.

The transformation of plagioclase minerals and an increased amorphous content would seem to be most related to the reduction in emanation power for Durango tailings. Decreasing the surface area during these transformations may be a prime factor in reducing emanation.

Nine composite materials were sintered at 1200°C in graphite and fire clay crucibles to produce relatively reducing versus oxidizing conditions. These tailings cover a wide range of composition and mineralogy; the emanating ^{226}Ra of these materials was reduced by factors of 22 to 1400 with final emanating ^{226}Ra values of 0.12 to 1.72 pCi/g. There were no great differences in emanation reduction between samples sintered in oxidizing and reducing atmospheres except for the Salt Lake City ferrophos and Durango fines which had lower emanating ^{226}Ra under reducing sintering conditions.

The large reduction in the emanating power, as well as the anticipated resistance to aqueous leaching, of thermally stabilized tailings implies that sintering of uranium mill tailings could produce a slag or clinker that could be disposed of on-site. The only cover required for a pile of thermally stabilized tailings would be a soil or rock layer to attenuate the gamma radiation field and to act as a substrate for vegetation. The engineering and economic considerations of thermal stabilization are discussed in a following section. Additional knowledge will be required before thermal stabilization can be applied as a remedial action technology:

(1)   the long-term stability of the sintered material exposed to physical degradation and chemical attack (e.g., solubility of minerals and amorphous material found in thermally stabilized tailings);

(2)   the interaction of the tailings and the refractories lining a kiln or other heating vessel;

(3)   the gaseous and particulate emissions produced during sintering of tailings;

(4)   revised engineering and economic analysis of the thermal stabilization process as more complete technical information is developed.

Despite these unknowns, the results of this laboratory study of thermal stabilization indicate that this approach to tailings management is technically feasible. The advantages of on-site disposal and the independence from barrier systems should decrease the complexity of remedial action and provide for the effective, long-term management of tailings.

SULFURIC ACID LEACHING FOR RADIONUCLIDE REMOVAL AND MINERAL RECOVERY

Another approach to remedial action for uranium mill tailings involves the removal of the components that are responsible for the environmental concern (notably ^{222}Rn releases) posed by these materials. Removing mineral values at the same time could defray much of the cost. This section presents laboratory results on sulfuric acid leachings and their effectiveness in accomplishing these aims (for details see reference [8]).

Composite tailings from the Shiprock, Salt Lake City, and Durango sites have been leached with deionized H_2O and 0.001N, 0.01N, 0.1N, 1N, 10N, and concentrated H_2SO_4 at a solid-to-liquid mass ratio of 1 to 5. The amounts of extractable material have been evaluated for 20 metals and non-metals, 5 anions, and 5 radioactive species.

The extraction of tailings constituents with concentrated H_2SO_4 differed greatly from that found with aqueous acids from 10N to 0.001N. The most apparent differences are as follows:

(1)   barium is extracted only by concentrated H_2SO_4 and not by any of the aqueous acids; from 30 to 70% of the Ba is extracted;

(2)   calcium is more extractable in conc. H_2SO_4 than for any aqueous acid;

(3)   metals such as Fe, Cu, Mn, Al, Mo, Ni, Pb, U, and V are not extracted as well by conc. H_2SO_4 as by aqueous acids with a final pH<3;

(4)   selenium is extracted most efficiently by conc. H_2SO_4;

(5)   conc. H_2SO_4 is more effective than 10N aqueous acid in extracting ^{226}Ra (70-80%), ^{230}Th (80-90%), and ^{210}Pb (20-55%).

The bahavior described above seems consistent with the proposition that conc. H_2SO_4 is an effective solvent of aqueously insoluble sulfate salts such as $BaSO_4$, $RaSO_4$, and $PbSO_4$ (or their hydrated forms). Those elements requiring free hydrogen ions to be solubilized, such as many of the transition metals, are not extracted as well by conc. H_2SO_4 (little free H^+ or SO_4^{-2}) as by strong aqueous acids.

When examining extractability by aqueous acids the importance of the pH of the leaching solution becomes apparent as well as the mineral form of element. Iron is somewhat extractable by 10N acid (4-33%) with lesser extraction occurring with 1N acid (1-16%). As expected, little soluble iron was found in leaching solutions with final pH's greater than 3. Similar trends are found with Cu, Co, Ni, Zn, and Al. Table IV shows extraction as influenced by acid concentration and final pH for Durango fines. All of the metals except U and V were insoluble in alkaline solutions; however, they are much less soluble in alkaline solutions than in acid solutions. These elements would be expected to form anion complexes with sulfate [9].

Analyses of anions in the tailings extracts indicates that phosphate minerals are being solubilized by strong aqueous H_2SO_4 in the Salt Lake City ferrophos and the Durango tailings. The small amounts of aluminum extracted (≤3%) from Shiprock and Salt Lake City tailings would indicate little breakdown of clays or feldspars. Durango tailings show somewhat greater aluminum extraction (6-9%). Probably, there is not a great deal of

Table V

Recoverable Mineral Values From Tailings Composites

Maximum Amount Extracted (g/metric ton)
Value Extracted ($/metric ton)

Mineral Value	Shiprock		Salt Lake City			Durango	
	Sands	Fines	Sands	Fines	Ferrophos	Sands	Fines
U	5	18	57	130	23	190	430
	*	*	$2.51	$5.72	$1.01	$8.36	$18.92
V	160	680	18	70	1560	450	2140
	*	$3.89	*	*	$8.92	$2.57	$12.24
Mo	43	130	120	1710	150	4	14
	*	$1.79	$1.65	$23.51	$2.06	*	*
Co	1	1	11	8	14	8	75
	*	*	*	*	*	*	$4.45
Cr	1	3	2	82	260	10	37
	*	*	*	*	$2.43	*	*
Mn	5	19	23	50	1340	160	680
	*	*	*	*	$2.06	*	$1.04
Ni	1	8	5	9	370	3	32
	*	*	*	*	$2.85	*	*
Cu	11	16	17	140	600	85	650
	*	*	*	*	$1.10	*	$1.20
Zn	5	13	25	72	33	180	1320
	*	*	*	*	*	*	$1.34
Total Value ($/metric ton)	*	$6.00	$4.00	$29.00	$20.00	$11.00	$39.00

*Value less than $1.00

solubilization of minerals originally present in the ore; minerals formed during or after milling are probably contributing much of the soluble components.

An assessment of mineral value recovery from these tailings is summarized in Table V. It is apparent that Shiprock tailings contain negligible values; whereas, the Salt Lake City fines and ferrophos residues and the Durango tailings contain appreciable values. These results indicate that a more thorough investigation of mineral recovery from these favorable tailings is warranted.

The results of our radionuclide extraction trials indicate that ^{226}Ra removal will require multiple stage leaching with concentrated sulfuric acid to achieve low residual radium in residues. Thorium-230 shows some evidence of being extracted by strong aqueous sulfuric acid, 10N. It is evident that removal of both of these precursors to ^{222}Rn is mandatory if such a vigorous approach to tailings conditioning is to be worthwhile.

TECHNICO-ECONOMIC ANALYSIS OF TAILINGS CONDITIONING

If a conditioning alternative is to be a viable alternative to some purely physical (barrier) method such as liners or covers, or is to be used in conjunction with a physical method, it must meet two conditions. First, it must be amenable to application of existing technology on a large scale with only a minimal time to adapt the technology. Second, the projected remedial action costs involved must be reasonably comparable with those of "equally effective" alternatives. (Details of this technico-economic analysis can be found in reference [10].)

Costs of any remedial action strategy can not be generalized across different tailings sites. Local factors make the cost of each alternative differ greatly between sites. Such site-specific factors mean that the economically desirable alternative at one site may be unacceptable at another.

The conceptual engineering design of a thermal stabilization process is based on the purchase of a used (coal-fired) rotary cement kiln, moving it to the site, installing appropriate feed apparatus and air-emissions control equipment, and repiling the processed "clinkers" on site with 0.6 m of gravel cover. The choice of coarse gravel cover was used to provide uniform comparison. Earth cover plus vegetative stabilization will be preferable at many sites.

Cost estimates for thermal stabilization of inactive tailings piles at Salt Lake City, Shiprock, and Canonsburg, Pennsylvania were calculated and the results summarized in Table VI. In comparing the costs it is seen that thermal stabilization, as in the conceptual process, is quite energy intensive. Therefore, at Shiprock, where both coal and electricity costs are quite low, the cost per ton for thermal stabilization is about the same as moving the pile a relatively short distance, but more than covering the same pile in place with earth. The thermal stabilization option has a relatively high labor cost, but, if applied in place as contemplated, certain other, very expensive labor costs are avoided. These are the costs of environmental impact studies, radiological monitoring in transport, and site de-contamination, which are incurred if the pile is to be moved from its original location. Thus, while the per ton cost for a small (but presumably highly radioactive) pile like Canonsburg appears quite high, it may be no more expensive than moving the pile through a densely populated area. Thus, the apparently high cost for thermal stabilization per ton of tailings may not be as high as other options when all costs are considered.

The more energy-efficient method of using combustible wastes for themal sta-bilization is within the realm of technological feasibility. Incineration with municipal wastes and hazardous organic chemical wastes are two methods of reducing fossil fuel usage which have been suggested. In addition, the kiln could be used as an incinerator after the completion of the thermal stabilization of tailings. Such a facility might be leased or sold to an appropriate local or state agency.

The general criterion for conducting leaching for mineral recovery as a part of a remedial action program requires that the economic benefits be at least equal to the incremental cost of adding the leaching step to the design scheme. A more specific pro-blem is whether it is economically feasible to recover mineral values from tailings as part of a remedial action program. The preliminary answer is yes for the Durango site, as explained in the following paragraphs.

Table VI

Costs of Thermal Stabilization of Uranium Mill Tailings

Item	Salt Lake City 450 MT/day ($)	Shiprock 450 MT/day ($)	Shiprock 900 MT/day ($)	Canonsburg 270 MT/day ($)
1. Capital Expense[a]	1.58	2.44	4.37	17.63
2. Direct Labor and Benefits	6.03	5.43	3.81	6.03
3. Coal (delivered)	5.10	1.53	1.53	6.58
4. Electricity[b]	0.79	0.62	0.62	1.10
5. Miscellaneous Operating Expenses	1.49	1.40	1.59	1.40
6. Flue Gas Desulfurization (if required)	(1.43)[e]	(1.43)[e]	(1.43)[e]	(2.20)[e]
7. Windblown and Off-site Remedial Action	5.68	f	f	f
8. Contractor Indirect[c]	3.97	2.05	1.62	2.23
9. Estimated Total and Direct	24.64	13.46	13.54	34.97
10. Contingency[d]	7.39	4.04	4.06	10.49
11. Grand Total	32.03	17.50	17.60	45.46
Total Costs For Thermal Stabilization	76,000,000	27,000,000	27,000,000	8,250,000

[a] Used kiln, auxiliaries, dust collection conveyors, including engineering and contractor charges. Possible salvage value not included.
[b] Salt Lake City, $0.05 KWH; Shiprock, $0.04 KWH; Canonsburg, $0.07 KWH.
[c] 30% of items 2, 5, and 7
[d] 30% of item 9
[e] Not included in total costs
[f] Not calculated for this site

In the late 1970s, only two sites of the locations available for study by Ranchers Exploration and Development Corporation (REDC) were deemed worthy of mineral recovery; these were Durango and Naturita, Colorado [11]. REDC was convinced that it could transport the Durango tailings to the Long Hollow Site, process them in excavated "leach tanks," cover with 0.6 m of compacted Mancos shale and make a profit. Based on 1978 prices and our current analysis and recovery data, the authors have estimated that REDC expected to gross at least $40,000,000 from the venture. Considering the risks involved, a profit margin of about $8,000,000 was probably used by REDC in 1978. This leaves a "production cost" of $32,000,000, which includes both leaching costs and hauling–excavation costs.

An independent engineering firm has reported the various costs of a hypothetical 2,000,000 ton, heap-leaching plant in southern Colorado or northern New Mexico [12]. Interpreting their figures liberally (high) for the Long Hollow disposal site and 1,600,000 tons of Durango tailings, yields a 1981 leaching cost of $20,800,000 from which credit can be taken for about $1,300,000 of health physics work that would have to be done as part of remedial action. Thus, the net leaching cost is $19,500,000.

Using current prices and our analytical data, the uranium and vanadium extracted from the 1,600,000 tons of Durango tailings would be worth $32,200,000, after discounting for the cost of off-site solution processing subsequent to heap leaching. The difference between the $19,500,000 leaching cost and the $32,200,000 mineral value would produce $12,700,000 in contribution margin to be applied to other costs of remedial action. If the only other costs are those listed by Ford Bacon and Davis Utah, Inc. in their Durango report [13], then the following advantages apply:

- Leaching would pay for 58% of the $21,800,000 cost of remedial action (other than leaching) with disposal at Long Hollow Site.

- The net cost would be less than that of reshaping, stabilizing, and covering the pile in place.

- Leaching and disposal at Long Hollow Site would be twelve million dollars less expensive than conventional off-site disposal.

This latter option is one that would make leaching and recovery of the strategic mineral values very difficult, if not impossible, both now or at some future time.

The high dollar and energy costs of earth moving mean that strategic minerals that may be recoverable from various tailings piles must either be recovered as part of the conditioning process or that disposal be carried out in a manner that will facilitate later recovery.

Whether sulfuric acid leaching of the tailings is conducted as part of an immediate uranium mill tailings remedial action program or is deferred to a later time will depend on the market prices for the mix of heavy metals in any given pile during the active remedial action phase. The technology for the leaching process and subsequent separations of uranium and vanadium is all available and well known, requiring only a relatively modest amount of applications research and development to adapt to the characteristics of any given tailings. However, technology for cobalt and other strategic mineral recovery needs to be developed. Any decision to use or not to use a leaching process should be determined on a basis of incremental benefit to the remedial action program and total recovery of strategic heavy metals. In other words, if the value of the materials recovered equals or exceeds the incremental cost of adding the leaching step, then leaching should be performed as part of tailings treatment. All strategic metals present in substantial concentrations in the acid leachates should be separated and recovered if the technology is available or can be quickly developed.

SUMMARY

A conditioning process may alter (1) the general chemical environment (pH, TDS, redox potential), (2) the concentrations of contaminants present in the conditioned tailings, (3) the general physical characteristics (particle size, surface area), or (4) the structure of the tailings (new mineral forms, crystalline vs amorphous). Evaluating changes in elemental mobility will help to clarify how the conditioned tailings have been modified and whether the product will retain advantageous characteristics over a long period. Understanding the chemical composition and mineralogical structure of uranium mill tailings in order to comprehend the physico-chemical alterations caused by conditioning often permits the conditioning process to be improved. Examining the reduction in emanating ^{226}Ra or emanation coefficients allows the control of ^{222}Rn flux resulting from the physico-chemical alteration of tailings to be evaluated on a laboratory scale.

Sintering tailings at high temperatures (1200°C) has been found promising as a conditioning approach that greatly reduces the ^{222}Rn emanation of uranium mill tailings. The structure of thermally stabilized tailings has been appreciably altered producing a material that will have minimal management requirements and will be applicable to on-site processing and disposal, assuming future alterations, if any, during weathering do not result in increased emissions.

Thermal stabilization (sintering) is somewhat expensive but could be economically attractive at tailings sites where some or all of the following conditions prevail:
1. Coal for kiln operation is inexpensive.
2. Topsoil for cover is not readily available.
3. Transportation costs to remote disposal area are high.
4. Environmental (radiological) monitoring costs are high for transport to remote disposal area.
The Shiprock and Canonsburg sites are believed to be economically attractive locations for thermal stabilization.

The technical feasibility of leaching hazardous radionuclides (namely ^{226}Ra, U, and ^{230}Th) and valuable metals from tailings with sulfuric acid has been demonstrated. Concentrated sulfuric acid is needed for the removal of the ^{226}Ra and strong aqueous acid is needed to leach the metal values. Multiple extraction would be needed to reduce ^{226}Ra, and hence ^{222}Rn, levels to acceptable values. Salt Lake City and Durango tailings contain substantial mineral values while the Shiprock tailings contain minimal recoverable values. Leaching, prior to disposal, is economically attractive wherever the value of recoverable strategic minerals exceeds the cost of conducting the leaching operation. The Durango and Salt Lake City tailings are sufficiently rich in uranium, vanadium or molybdenum to make a leaching treatment very desirable.

REFERENCES

1. US Environmental Protection Agency: "Draft Environmental Impact Statement for Remedial Action Standards for Inactive Uranium Processing Sites (40 CFR 192)" EPA 520/4-80-011, December 1980.

2. Cokal, E. J. et al.: "The Chemical Characterization and Hazard Assessment of Uranium Mill Tailings," Proc. Symp on Uranium Mill Tailings Management, pp. 4763, Colorado State University, Fort Collins, Colorado, 1981.

3. Dreesen, D. R. et al.: "Thermal Stabilization of Uranium Mill Tailings," Proc. Symp on Uranium Mill Tailings Management, pp. 65-79, Colorado State University, Fort Collins, Colorado, 1981.

4. US Environmental Protection Agency: "Proposed Criteria for Water Quality," Vol. 1, October 1973.

5. US Environmental Protection Agency: "National Interim Primary Drinking Water Regulations," EPA-570/9-76-003, 1976.

6. US Environmental Protection Agency: "Proposed Disposal Standards for Inactive Uranium Processing Sites; Proposed Rule and Extension of Comment Period," Federal Register, January 9, 1981.

7. Cleland, J. and G. L. Kingsbury: "Multimedia Environmental Goals for Environmental Assessment," Vol. 1, US Environmental Protection Agency report EPA-600/7-77-136a, November 1977.

8. Williams, J. M. et al.: "Removal of Radioactivity and Mineral Values from Uranium Mill Tailings," Proc. Symp. on Uranium Mill Tailings Management, pp. 81-95, Colorado State University, Fort Collins, Colorado, 1981.

9.   Merritt, R. C.: The Extractive Metallurgy of Uranium, p 576, Johnson Publishing Co, Boulder, Colorado, 1971.

10. Thode, E. F. and D. R. Dreesen: "Technico-Economic Analysis of Uranium Mill Tailings Conditioning Alternatives," Proc. Symp. on Uranium Mill Tailings Management, pp. 155-165, Colorado State University, Fort Collins, Colorado, 1981.

11. Deuel, J. K.: "The Durango Mill Tailings Story: Everyone Loses," Proc. Symp. on Uranium Mill Tailings Management, pp. 293-301, Colorado State University, Fort Collins, Colorado, 1979.

12. Mountain States Research and Development Corp.: Uranium Mill Tailings Remedial Action Project report to Sandia National Laboratories, Albuquerque, New Mexico, 1980.

13. Ford, Bacon, and Davis, Utah Inc.: "Engineering Assessment of Inactive Uranium Mill Tailings - Durango Site, Durango, Colorado," DOE/UMT-0103, June 1981.

14. Bowen, H. J. M.: Environmental Chemistry of the Elements, p 333, Academic Press, London, 1979.

URANIUM TAILING DISPOSAL
BY THE THICKENED TAILING DISCHARGE SYSTEM

E. I. Robinsky

Professor of Civil Engineering, University of Toronto
and
President, E. I. Robinsky Associates Limited, 66 Lytton Blvd.,
Toronto, Canada, M4R 1L3

Synopsis

It is shown that the Thickened Tailing Disposal System results in less environmental impact than most other systems used in the mining industry today. The initial and operating costs are generally lower and the system can be adapted to any given topography. The most damaging aspect of tailing disposal is usually seepage into the surronding environment. Such seepage cannot be prevented entirely, but the thickened tailing disposal system creates less contamination than conventional methods. It maximizes evaporation, optimizes runoff and eliminates dusting. The construction of high and costly confining dikes and dams is eliminated. Finally, it is shown that the tailing disposal area can be readily rehabilitated. One case study is given to show how topographical features can be used to advantage to dispose of tailing safely and economically, using the thickened tailing disposal system.

Introduction

The Thickened Tailing Disposal System has only recently been introduced (Robinsky, 1979) into the mining and mineral processing industry. This paper describes the system and indicates its principal merits. The environmental impact is discussed and the use of natural topographical features is illustrated by one case study.

Principles of the System

The system consists of thickening the tailing and discharging the slurry through permanently-established spigots overlooking the disposal area. The tailing, being of a relatively thick consistency, will come to rest beneath the discharge spigots at its natural angle of reponse forming a conical side-slope against the valley wall or discharge ramp. By judicious positioning of the spigots it is possible to regulate the configuration and elevation of the tailing deposit, while at the same time eliminating a major environmental hazard -- the need for confining perimeter dikes. If the area already contains dikes, the use of the thickened discharge system will make it possible to dispose of considerably more tailing without increasing the height of the dikes.

The system also eliminates the large flat 'slimes' pond -- a normal feature of conventional disposal operations. A tailing which has been thickened to slurry consistency will not segregate during deposition. The fine-grained fraction (slimes) of the tailing remains disseminated evenly throughout the deposit. This is contrary to the behavior of unthickened conventionally discharged tailing where the coarse-grained fraction settles out close to the spigots and the fines are carried out to the 'slimes' pond.

Although there are no fixed limits regarding the tailing slope that can be chosen, a six percent slope (3.4°) has been commonly recommended as most practical. Reclamation and climatic conditions generally govern this decision and the slope may be adjusted by adjusting the thickening of the tailing.

The tailing deposit is accompanied by a small tailing water pond located generally within the limits of the disposal area. Sometimes, however, the pond may be established outside the limits in order to provide more room for tailing. By design this pond receives both tailing and natural runoff from the tailing deposit and surrounding watershed, together with a small fraction of tailing fines transported by the runoff. All runoff water eventually reaches the tailing water pond naturally, without constructing special drainage systems. Natural terrain contours are used to advantage for this purpose, together with designed positioning of the discharge spigots.

Figure 1   Comparison of the conventional and thickened tailing disposal systems

Depositional Behavior of Thickened Tailing

Generally, it would be anticipated that even with thickened tailing, the coarser fraction should settle out closer to the point of discharge in accordance with the laws of sedimentation. This segregation by size, however, does not occur. The heavy slurry flows downhill from the spigots without effective sorting. When the slurry reaches an underlying slope less than the natural depositional slope of the thickened tailing, the viscosity of the material overcomes the gravitational and dynamic forces acting on the flowing slurry, and the flow slows to a stop. Thus, layer by layer the deposit accumulates. Because any liquid flow will always seek the steepest slope and spread out on flat surfaces, the deposit will build up in thin layers of very uniform thickness, generally less than 2 inches (50mm). Because there is no segregation of materials, the viscosity remains constant whether the tailing is near the discharge spigot or several thousand feet away. Therefore the slope attained anywhere along the deposit is identical and uniform.

There are several factors which determine the steepness of the slope of the deposited tailing, viz.:

a. Percent solids of the tailing at discharge. As a tailing is thickened it progressively changes from a mixture of solid particles and liquid to a non-segregating slurry. The state at which the latter occurs is the minimum percent solids to which any tailing must be thickened. Typically, a slurry of this consistency will stand at slopes of 1 to 2 percent. Additional thickening will result in steepening of the deposited slope at an ever-increasing rate. Figure 2 illustrates typical laboratory tailing deposition test results.

b. Grain-size distribution of the tailing. The finer the tailing, the less percent solids is required to achieve the same depositional slope. The coarser the tailing, the more thickening is necessary. Figure 3 shows the results of tests made on many different tailing samples. For every mine a different relation is obtained between the slope of the deposited tailing and its percent solids.

c. The pH of the tailing. The pH of the tailing can have a very significant effect on the depositional slope. This appears to apply only to tailings having a pH on the basic side, but not necessarily to all tailing types.

d. Rate of discharge per spigot. Another important factor which affects the slope of the deposition is the volume of flow from the discharge point. It has been found that if the same volume of material is discharged from several points instead of one discharge point, the resultant slope is steepened by as much as 0.75 to 1.0 percent. In order to benefit from this phenomenon and thereby reduce the degree of thickening to a minimum, it is advantageous to install a series of spigots for simultaneous discharge.

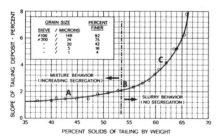

Figure 2   Typical laboratory tailing
deposition test results

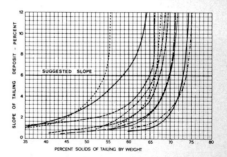

Figure 3   Laboratory test results showing
tailing deposition curves.
Tailing samples were obtained
from different mines

The depositional behavior of the tailing and the desired final tailing slope will dictate the tailing disposal program. Initially, the percent solids could be at the point of commencement of 'slurry behavior' (see point B in Figure 2). If tailing is being discharged into a valley, the initial percent solids is maintained until the toe of the resulting slope reaches the property line or tailing area boundary. At this time the percent solids of the tailing being produced is increased (say, to point C in Figure 2). This produces a steeper slope with no change in the position of the toe. If discharge is from a built-up ramp within the disposal area, the discharge spigots and ramp will also have to be raised at this time. The requirement of progressive thickening may permit the gradual acquisition of thickening equipment. Thus a minimum number of thickening units may be required at the start, followed by additional units being installed in subsequent years, as dictated by the tailing disposal schedule.

As the tailing deposit increases in thickness the underlying tailing consolidates to an ever-increasing density. Even small increases in percent solids create substantial increases in the strength of the deposit, as represented by the rapidly increasing slope of the deposition curve, Figure 2.

Environmental Concerns

Selection of Disposal Site

There is a strong natural tendency in the industry to search out a location for tailing disposal as close as possible to the mill, even when better disposal areas are available at some distance. This tendency has been inherited from the not-so-distant past when environmental concerns did not exist. The presence of a natural adjacent valley or trough has been, and still is, considered fortuitous. The prime consideration was to reduce immediate costs of pipeline installation, pumping and maintenance. However, recent technical developments and imposition of environmental controls have caused a reconsideration of this approach. Disposal at a more remote but environmentally more acceptable site may now be more economical.

The use of the thickened tailing disposal system allows a much wider selection of sites to be considered. The sites could be flat, hilly or mountainous. The system can be adopted to a variety of topographical settings. Containment dikes can be small or even non-existant, as illustrated by the case study described herein. Distance to the disposal area should not prevent consideration of alternate sites. Pumping technology has come a long way in the past decade. Abrasion-resistant pump linings have been developed and pump maintenance costs have been sharply reduced. To control the impact of a pipeline spill, automatic shut-down valves, spillage sumps and concentric double pipelines should be considered. Furthermore, the impact of a spill of thickened tailing slurry is less severe than a spill of unthickened highly liquid tailing. Finally, pumped as a slurry, the tailing will require substantially smaller pipelines and less pumping energy. The mass of tailing to be pumped is generally less than one-half of the unthickened material. The volume of return water is likewise reduced to approximately one-fifth of that of conventional disposal systems.

Dusting

Dusting develops along the dry perimeter surfaces and beaches of conventionally discharged tailing. At low percent solids in conventional disposal operations the tailing segregates into sand, silt and clay fractions as it flows from a discharge spigot. It is the fine sand and silt-size fraction, left on the dry perimeter beach, that is susceptible to being picked up and transported by wind. Conventional tailing disposal operations require periodic repositioning of the discharge lines and spigots. After the spigots are moved, the area dries out and dusting can result. With thickened tailing disposal dusting is prevented for the following reasons.

a.   The discharge spigots remain essentially at the same location until termination of operations. This ensures that the tailing surfaces are continually rewetted.

b.   Because there is no segregation of tailing particles after discharge, the capillary rise in the tailing is determined everywhere by the finest fraction -- the clay size. Such material has very high capillary rise. Thus there is a continuous rise of pore fluid to the surface of the tailing which tends to keep it moist.

c.   It is envisaged that after completion of the mining operations, the surface of the tailing will become desiccated to produce a dense dust-free surface. This occurs because of the even dispersal of fine clay-size particles which act as a bonding agent throughout the tailing. Such behavior has been observed generally with all tailings that have been tested, including uranium tailings.

It is thus concluded that dusting should not be a problem if the thickened tailing disposal system is used. After termination of operations, the reclamation procedures suggested herein can be used to create an effective earth cover.

Resistance to Erosion by Runoff

During operations some erosion of the tailing by runoff from precipitation is anticipated and provision must be made to accommodate the eroded tailing. One way is to provide space in the runoff pond for 2 or 3 percent of the total volume of tailing. Alternatively, a better approach is to maintain the toe of the developing deposit somewhat short of the tailing disposal area boundary. The runoff material will settle in this area and eventually will be covered by tailing discharged during the last year or two of operations.

Runoff velocity is the primary cause of surface erosion. Velocity is dependent on the slope and on the thickness of the flowing sheet of water. Climatic conditions and the length of the tailing surface will determine these factors.

Three mechanisms are identified that act simultaneously either to decrease or increase erosion potential. Because of the complexity of the problem the interaction of these mechanisms is difficult to predict.

a. One of the favorable features is the retention of the fine particles (slimes) in the tailing slurry due to the process of thickening prior to discharge. The fine particles act as a binder for the coarse fraction due to inter-particle friction and cohesion. Figure 4 illustrates the critical water velocities (scour velocity) for quartz sediment -- a material not unlike a mine tailing. The curves show that particle sizes in the range of 0.1 to 0.5 mm (fine to medium sand) will scour at the lowest velocity. By discharging unsegregated tailing whose mean particle size is considerably smaller at 0.01 to 0.05 mm, a higher velocity for scour is necessary. On a tailing slope of say 4 percent a moderate amount of runoff theoretically will flow at a velocity of approximately 0.7 ft/sec. (200 mm/sec). It thus appears that many deposits would undergo only minor erosion, the developed velocity being too low to scour the surface.

b. An unfavorable mechanism that is expected to increase erosion is the impact of rain drops on the dry (or wet) surface of the tailing. Each drop will tend to pluck out some tailing particles. Once splashed out, the particles are carried away by sheet flow. In addition the rain drop impact creates turbulence within the sheet flow which, in turn, increases sheet erosion.

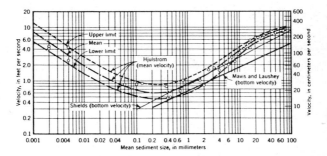

Figure 4   Critical (scour) water velocities for quartz sediment as function of mean grain size (ASCE, 1975)

c. The third mechanism in the erosion process is a beneficial one. The flowing sheet of water may wash out some of the fine and intermediate particles from between the coarser particles and storage capability for these eroded fines must be provided as stated earlier. However, as the upper finer particles are removed, the remaining coarser particles may create a protective blanket against further erosion.

On the basis of the equation developed by Smith and Wischmeier (1957, 1962) for estimating soil loss by all surface water erosion mechanisms, it is calculated that loss of tailing on slopes of 4 to 7 percent under very high rainfall conditions could amount from 60 to 100 tons per acre per year. For typical tailing disposal areas, these quantities represent 1.5 to 2.5 percent of the total tailing discharged. It is thus concluded that erosion of the tailing surface by precipitation and runoff is not serious for typical design slopes of 4 to 6 percent.

It is, of course, expected that during active mining operations the contaminated water in the runoff pond will be either recycled or treated before being released.

Stability under Earthquake Loading

Earthquake effects are considered to be minimized on a deposit of tailing at a 4 to 6 percent slope, discharged by the thickened tailing disposal method. Thickened tailing, when discharged, is in a completely liquified state. This state is similar to that produced by liquefaction of a sand or silt deposit during an earthquake. It is under this liquefied condition (steady state condition) that the tailing will flow until it reaches equilibrium with the underlying slope. After deposition the material consolidates (excess water is squeezed out), and its shearing strength increases very rapidly. The rate at which strength is gained can be measured by the slope obtained in laboratory deposition tests (see Figure 2). As the tailing consolidates and increases in strength it acquires a capability of standing in a liquefied condition at a much steeper slope than that at which it was deposited. The factor of safety against flow gradually continues to rise.

In the upper few feet of the deposit, where consolidation is still progressing rapidly, it is possible that some mass movement due to inertia during an earthquake could occur. At a depth of 3 to 4 feet, where the tailing is more consolidated, movement is restricted. Furthermore, no movement is expected to occur after the termination of the earthquake, because the material in a homogeneous artificially liquefied state (created by the earthquake) does not contain sufficient liquid to convert the material back to a flowable condition. A large amount of liquid is required to cause the tailing to flow again on the slope at which it was deposited, a slope of 4 to 6 percent.

Seepage

Disposal by the thickened tailing discharge system has considerable advantages over the conventional approach with regard to reducing the amount of seepage into the underlying soil formations, both during operations and after reclamation.

a. In the first place, because the tailing has been thickened to slurry consistency, it will contain, at discharge, approximately one-half of the mass of water it would have contained in the unthickened state. Therefore a smaller quantity of liquid has to be handled in the tailing area and runoff pond.

b. The thickened tailing deposit is not contained by dikes, or has much smaller dikes and therefore seepage through such structures is either eliminated, or at least greatly reduced.

c. By the elimination of the slimes pond the free hydrostatic head which is found on conventional disposal areas is also eliminated. Without a free head of liquid above the sloped tailing surface, any downward seepage movement will be immediately arrested by the development of a substantial capillary rise potential (suction). This will occur because the fines in the tailing are disseminated throughout the mass and it is the fine fraction which contributes most to the capillary forces. Capillary rise in the typical fine fraction of mine tailing can theoretically attain 150 or more feet.

d. Oxidation of sulphides present within a mass of tailing can generate acid which, in turn, may cause the dissolution of heavy metals. This condition does not arise in the thickened tailing disposal system. The slope of the tailing surface and the continually saturated condition of the deposit during its active life should prevent the penetration of any oxygen-containing precipitation into the the tailing. The tailing should remain essentially unaltered. After reclamation, the sloping surface still provides a major advantage in that runoff will occur rapidly. Climatic conditions and the surface treatment will determine the amount of precipitation that will be absorbed.

In summary, ground pollution by seepage is considerably diminished and can even be eliminated under certain conditions, as in the case study described in this paper.

Radon Emanation and Radiation

There are several factors provided by the thickened tailing disposal system that should reduce radon emanation and radiation.

a. Radium tends to concentrate in the 'fines' of uranium tailing. Therefore, in conventional disposal basins the 'slimes' pond tends to be a high radiation area. With the thickened tailing disposal system where the tailing is a homogeneous non-segregated slurry, areas of concentrated radiation should not arise.

b. Radon gas will diffuse more readily in a dry porous mass than in a saturated dense one. Because the fines in thickened tailing are disseminated uniformly throughout the voids of the coarser fraction, the material is more dense. In addition, the fines provide water-retention capability and high capillary rise potential, both acting to maintain a saturated dense condition.

c. Finally, normal operating procedures will require that the discharge spigots remain essentially at the same location until reclamation is commenced. This ensures that the tailing surfaces are continually re-wetted, thus providing a saturated mass of tailing which should reduce radon diffusion.

Reclamation

One of the common requirements for successful reclamation of a tailing pond is to ensure that the final soil surface is not too wet nor too dry for plant growth. Such a condition is provided automatically by the thickened tailing disposal system. No regrading of surfaces is necessary. The gentle planned slope of 4 to 6 percent provides good drainage which also enhances surface evaporation. There is no slimes pond nor a buried decant pipeline as in conventional tailing ponds. The system contains only a simple overflow structure at the lowest topographical point in the disposal area. No major operation is involved in abandoning this structure.

With regard to surface treatment and re-vegetation of a tailing disposal area, the system provides a substantial advantage over conventional disposal areas. It is suggested that the surface can be treated and seeded immediately after termination of tailing deposition, without the aid of any earthmoving or farming equipment.

Some weeks or months prior to the termination of mining operations, appropriate neutralizer, fertilizer, seed and possibly topsoil is added to the tailing immediately prior to discharge. Because the tailing is in slurry form segregation of the added ingredients does not occur. Furthermore, the build-up of the material occurs in a layer of uniform thickness. The longer the discharge is maintained at one location, the thicker will be the deposited treated tailing. It would therefore be possible to place a uniformly treated layer of preplanned thickness and cover the entire tailing surface. The process is automatic. Furthermore, in most topographical settings, proper planning of the positions and elevations of the tailing discharge spigots will allow progressive reclamation procedures to take place. That is, it should be possible to progressively fill a tailing disposal area to its ultimate tailing height, and then to follow immediately with progressive reclamation.

Finally, if there is any question regarding the ability of the tailing to sustain vegetation, in spite of the application of neutralizer, an alternative solution is proposed. It also relies on the use of the thickened tailing disposal concept. Towards the end of mining operations natural rock and/or soil from the vicinity of the mine or from the upper benches of an open pit mine is to be used as a soil cover. Alternatively, 'clean' waste rock could be stockpiled for this future use. The material must be free of sulphides and other potentially toxic ingredients. It is to be mined, crushed, milled and thickened to form a slurry which can then be released through the existing tailing line onto any new or old tailing surface. By control of the slurry consistency it will be possible to match the slope of the soil cover to that of the underlying tailing, thereby assuring a uniform and consistent cover. It would no doubt be advantageous to add some fertilizer to the soil before discharge. Possibly shredded roots of local fast-growing plants and/or seeds could be added as well. It is expected, of course, that only the crushing and milling circuits would be in operation; the floatation circuit, etc. would be closed down. Thickeners would still be required.

As an example, a 9000 ton per day concentrator would produce enough material to cover a 200 acre tailing pond with 8 inches of non-toxic prepared soil (at 75 percent solids) in a period of only 28 days.

Case Study

A case study is described hereunder to illustrate the advantages of the thickened tailing disposal system. The project described is one of several alternative schemes proposed for the disposal of uranium tailing for Gulf Mineral Resources Company at their Mount Taylor project in New Mexico.

Figure 5 shows the general topography of the small dry valley chosen for the disposal of tailing for 17 years of production (20,000,000 tons). Also shown are the natural existing slopes, the limits of the watershed and the envisaged tailing limits. Because of the low annual precipitation rate (10 in., 250 mm) and high annual evaporation rate (57 in., 1448 mm) a relatively small evaporation pond is to be provided behind a temporary dam. The tailing is to be discharged to a 6 percent slope. It is suggested that the proposed deposit should satisfy the intent of the U.S. environmental regulation for complete burial. The deposit will be below any adjacent natural terrain. Furthermore, no dikes will be required to contain the tailing.

To satisfy another environmental regulation, that of providing a 10 ft (3 m) thick soil cover, it is proposed to excavate the disposal site and evaporation pond and stockpile the soil around the periphery of the disposal area for later use, as shown in Figure 6.

Discharge of tailing is to be effected by a number of spigots operating simultaneously and in accordance with the schedule shown on Figure 7. The position and elevation of the spigots is designed to provide complete control of the shape and elevation of the tailing deposit. The extent of the surface area of the discharged tailing is planned to provide drying capacity sufficient to attain continuously 80 percent solids in the tailing -- a fully consolidated state.

One of the advantages of the thickened tailing disposal system is that progressive reclamation is possible. Figure 8 shows a late stage in the disposal program. The North end has already been reclaimed.

Finally, Figure 9 shows the site fully reclaimed. It is to be noted that the toe of the tailing deposit has been kept well inside the mouth of the valley. Natural runoff will, for the most part, have to cross the original erosion-resistant crests of the chosen area (see Figure 5), providing thereby substantial protection for the deposited tailing.

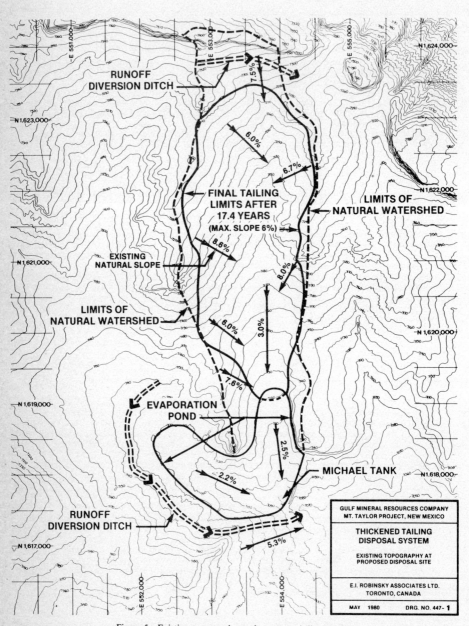

Figure 5   Existing topography at the proposed disposal site

Conclusions

In conclusion, the thickened tailing disposal system provides the following advantages over the conventional system.

Any chosen disposal area can accommodate considerably more tailing than could be accommodated by conventional storage systems.

Overall cost of tailing disposal is reduced substantially through the elimination of confining dikes, the reduction in size of tailing and return water pipelines, and the elimination of pipe abrasion by pumping at low velocity. Finally, reclamation procedures are simple and automatic.

The environmental impact is small. Dusting is inhibited and seepage potential greatly reduced by the elimination of the slimes pond and the provision of a self-draining sloped surface. Safety is enhanced through the elimination or reduction in size of the perimeter confining dikes.

Acknowledgement

The writer wishes to thank Gulf Mineral Resources Co. for allowing the use of material from a study made for their proposed operation at Mount Taylor, New Mexico.

References

Robinsky I. Eli, Tailing Disposal by the Thickened Discharge Method for Improved Economy and Environmental Control, Proceedings, 2nd International Tailing symposium, Colorado, Tailing Disposal Today, Volume 2, Miller Freeman Publications, Inc., San Francisco, California, 1979.

Sedimentation Engineering, ASCE, Manuals and Reports on Engineering Practice, No. 54, 1975, p.102.

Smith, D. D., and W. H. Wischmeier, Factors Affecting Sheet and Rill Erosion, Trans. Am. Geophys. Union 38, 1957, p. 889-896.

Smith, D. D., and W. H. Wischmeier, Rainfall Erosion, Advances in Agronomy 14, 1962, p. 109-148.

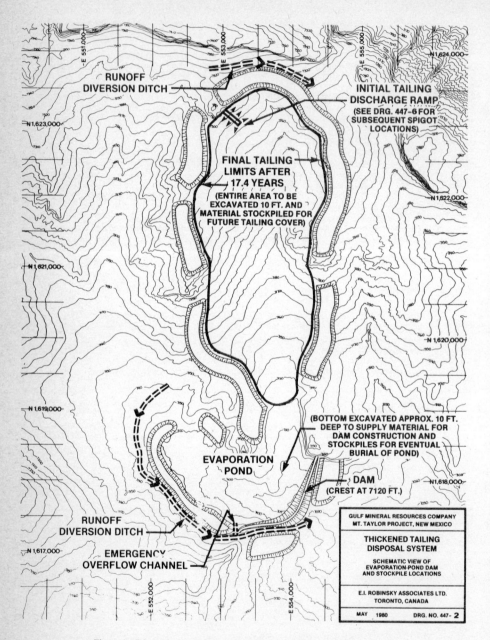

Figure 6 Schematic view of evaporation pond dam and stockpile locations

Within the figure:

RUNOFF DIVERSION DITCH

INITIAL TAILING DISCHARGE RAMP (SEE DRG. 447-6 FOR SUBSEQUENT SPIGOT LOCATIONS)

FINAL TAILING LIMITS AFTER 17.4 YEARS (ENTIRE AREA TO BE EXCAVATED 10 FT. AND MATERIAL STOCKPILED FOR FUTURE TAILING COVER)

(BOTTOM EXCAVATED APPROX. 10 FT. DEEP TO SUPPLY MATERIAL FOR DAM CONSTRUCTION AND STOCKPILES FOR EVENTUAL BURIAL OF POND)

EVAPORATION POND

DAM (CREST AT 7120 FT.)

RUNOFF DIVERSION DITCH

EMERGENCY OVERFLOW CHANNEL

GULF MINERAL RESOURCES COMPANY
MT. TAYLOR PROJECT, NEW MEXICO

THICKENED TAILING DISPOSAL SYSTEM

SCHEMATIC VIEW OF EVAPORATION-POND DAM AND STOCKPILE LOCATIONS

E.I. ROBINSKY ASSOCIATES LTD.
TORONTO, CANADA

MAY 1980   DRG. NO. 447-2

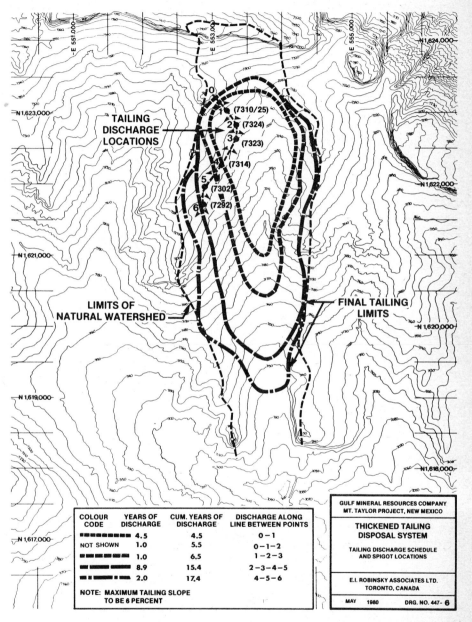

Figure 7 Tailing discharge schedule and spigot locations

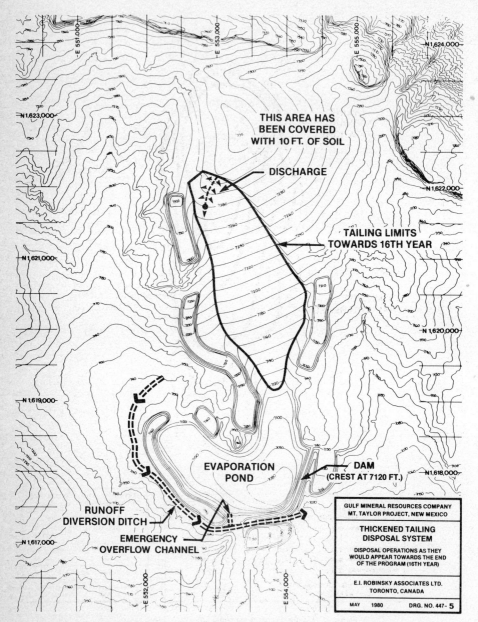

THIS AREA HAS
BEEN COVERED
WITH 10 FT. OF SOIL

DISCHARGE

TAILING LIMITS
TOWARDS 16TH YEAR

EVAPORATION
POND

DAM
(CREST AT 7120 FT.)

RUNOFF
DIVERSION DITCH

EMERGENCY
OVERFLOW CHANNEL

GULF MINERAL RESOURCES COMPANY
MT. TAYLOR PROJECT, NEW MEXICO

THICKENED TAILING
DISPOSAL SYSTEM

DISPOSAL OPERATIONS AS THEY
WOULD APPEAR TOWARDS THE END
OF THE PROGRAM (16TH YEAR)

E.I. ROBINSKY ASSOCIATES LTD.
TORONTO, CANADA

| MAY | 1980 | DRG. NO. 447- 5 |

Figure 8   Disposal operations as they would appear towards the end of the program (16th year)

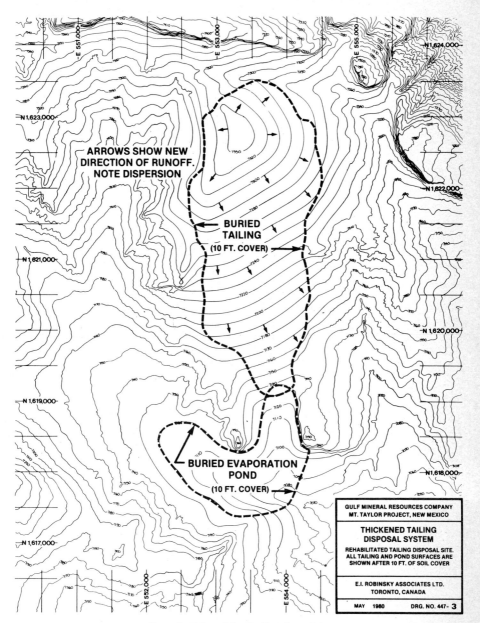

Figure 9   Fully reclaimed tailing disposal site

LIST OF PARTICIPANTS

LISTE DES PARTICIPANTS

AUSTRALIA - AUSTRALIE

LEVINS, D.M., Leader, Chemical Engineering Section, Australian Atomic
    Energy Commission, Research Establishment, Lucas Heights, NSW 2232

CANADA

BAIRD, M., Department of Chemical Engineering, McMaster University,
    Hamilton, Ontario L8S 4L7

BLAND, J., Department of Physics, University of Calgary,
    2500 University Drive N.W., Calgary, Alberta T2N 1N4

BRAGG, K., Senior Division Officer, Atomic Energy Control Board,
    270 Albert Street, Ottawa, Ontario K1P 5S9

CAMPBELL, M.C., Manager, Extraction Metallurgy Laboratory, CANMET,
    555 Booth Street, Ottawa, Ontario K1A 0G1

HOWIESON, J., Adviser, Nuclear Waste Management, Energy, Mines and
    Resources, 580 Booth Street, Ottawa, Ontario K1A 0E4

JOE, E.G., CANMET, Energy, Mines and Resources, 552 Booth Street,
    Ottawa, Ontario K1A 0G1

LENDRUM, F.C., Consulting Engineer, P.O. Box 70, King City,
    Ontario L0G 1K0

LUSH, D., Manager, Aquatic Ecologist, Beak Consultants Limited,
    6870 Goreway Drive, Mississauga, Ontario L4V 1L9

OSBORNE, R.V., Head, Environmental Research Branch, Atomic Energy of
    Canada Ltd., Chalk River Nuclear Laboratories, Chalk River,
    Ontario K0J 1J0

PULLEN, P.F., Consultant, 274 Riverside Drive, Oakville,
    Ontario L6K 3N4

ROBINSKY, E.I., Professor, University of Toronto and Consulting
    Engineer, 66 Lytton Boulevard, Toronto, Ontario M4R 1L3

VASUDEV, P., EPS, Place Vincent Massey, Environment Canada,
    Ottawa K1A 1C8

FEDERAL REPUBLIC OF GERMANY - RÉPUBLIQUE FÉDÉRALE D'ALLEMAGNE

SCHNEIDER, B., Urangesellschaft mbH, Bleichstrasse 60/62,
    D-6000 Frankfurt

SOUTH AFRICA - AFRIQUE DU SUD

SIMONSEN, H.A., National Institute for Metallurgy, Private Bag X3015,
    Randburg (2125)

UNITED STATES - ÉTATS-UNIS

BUSH, K.J., GEC Research, Inc., 2693 Commerce Road, Rapid City, SD 57701

DREESEN, D.R., Environmental Science Group, Los Alamos National Laboratory, MS 495, Los Alamos, New Mexico 87545

GONZALES, D.E., Environmental Consultant, P.O. Box 692, Nederland, Colorado 80466

LANDA, E., US Geological Survey, Mail Stop 424, Box 25046, Federal Center, Denver, Colorado 80225

MARKOS, G., GEC Research, Inc., 2693 Commerce Road, Rapid City, SD 57701

ROGERS, V.C., Chief Scientist, Rogers & Associates Engineering Corporation, 515 East 4500 South, Suite G-100, Box 330, Salt Lake City, Utah 84110

RYAN, A.D., ORNL, P.O. Box X, Oak Ridge National Laboratory, Oak Ridge, Tennessee 37830

SNODGRASS, W.J., Department of Geography and Environmental Engineering, 415 Amco Hall, Homewood Campus, John Hopkins University, Baltimore, Maryland 21218 /on sabbatical leave with Beak Consultants ; see address of Dr. D. Lush, Canada_7

INTERNATIONAL ATOMIC ENERGY AGENCY
AGENCE INTERNATIONALE DE L'ÉNERGIE ATOMIQUE

SEIDEL, D.C., Division of Nuclear Fuel Cycle, International Atomic Energy Agency, P.O. Box 100, A-1400 Vienna, Austria

OECD NUCLEAR ENERGY AGENCY
AGENCE DE L'OCDE POUR L'ÉNERGIE NUCLÉAIRE

TAYLOR, D.M., Nuclear Development Division, OECD Nuclear Energy Agency, 38 boulevard Suchet, 75016 Paris, France

SOME
NEW PUBLICATIONS
OF NEA

QUELQUES
NOUVELLES PUBLICATIONS
DE L'AEN

ACTIVITY REPORTS

RAPPORTS D'ACTIVITÉ

Activity Reports of the OECD Nuclear Energy Agency (NEA)
— 8th Activity Report (1979)
— 9th Activity Report (1980)

Rapports d'activité de l'Agence de l'OCDE pour l'Énergie Nucléaire (AEN)
— 8e Rapport d'Activité (1979)
— 9e Rapport d'Activité (1980)

Free on request — Gratuits sur demande

Annual Reports of the OECD HALDEN Reactor Project
— 19th Annual Report (1978)
— 20th Annual Report (1979)

Rapports annuels du Projet OCDE de réacteur de HALDEN
— 19e Rapport annuel (1978)
— 20e Rapport annuel (1979)

Free on request — Gratuits sur demande

• • •

INFORMATION BROCHURES

BROCHURES D'INFORMATION

— OECD Nuclear Energy Agency: Functions and Main Activities

— NEA at a Glance
— International Co-operation for Safe Nuclear Power
— The NEA Data Bank

— Agence de l'OCDE pour l'Énergie Nucléaire : Rôle et principales activités
— Coup d'œil sur l'AEN
— Une coopération internationale pour une énergie nucléaire sûre
— La Banque de Données de l'AEN

Free on request — Gratuits sur demande

• • •

SCIENTIFIC AND TECHNICAL PUBLICATIONS

PUBLICATIONS SCIENTIFIQUES ET TECHNIQUES

NUCLEAR FUEL CYCLE

LE CYCLE DU COMBUSTIBLE NUCLÉAIRE

World Uranium Potential —
An International Evaluation (1978)

£7.80   US$16.00   F64.00

Potentiel mondial en uranium —
Une évaluation internationale (1978)

Uranium — Ressources, Production and Demand (1982)

£9.90   US$22.00   F99,00

Uranium — ressources, production et demande (1982)

Nuclear Energy and Its Fuel Cycle: Prospects to 2025 (in preparation)

L'énergie nucléaire et son cycle de combustible : perspectives jusqu'en 2025 (en préparation)

• • •

SCIENTIFIC INFORMATION

INFORMATION SCIENTIFIQUE

Neutron Physics and Nuclear Data for Reactors and other Applied Purposes
(Proceedings of the Harwell International Conference, 1978)

£26.80   US$55.00   F220,00

La physique neutronique et les données nucléaires pour les réacteurs et autres applications
(Compte rendu de la Conférence Internationale de Harwell, 1978)

Calculation of 3-Dimensional Rating Distributions in Operating Reactors
(Proceedings of the Paris Specialists' Meeting, 1979)

£9.60   US$21.50   F86.00

Calcul des distributions tridimensionnelles de densité de puissance dans les réacteurs en cours d'exploitation (Compte rendu de la Réunion de spécialistes de Paris, 1979)

Nuclear Data and Benchmarks for Reactor Shielding
(Proceedings of a Specialists' Meeting, Paris, 1980)

£9.60   US$24.00   F96,00

Données nucléaires et expériences repères en matière de protection des réacteurs
(Compte rendu d'une réunion de spécialistes, Paris, 1980)

• • •

RADIATION PROTECTION   RADIOPROTECTION

Iodine-129
(Proceedings of an NEA Specialist Meeting, Paris, 1977)

Iode-129
(Compte rendu d'une réunion de spécialistes de l'AEN, Paris, 1977)

£3.40   US$7.00   F28,00

Recommendations for Ionization Chamber Smoke Detectors in Implementation of Radiation Protection Standards (1977)

Recommandations relatives aux détecteurs de fumée à chambre d'ionisation en application des normes de radioprotection (1977)

Free on request — Gratuit sur demande

Radon Monitoring
(Proceedings of the NEA Specialist Meeting, Paris, 1978)

Surveillance du radon
(Compte rendu d'une réunion de spécialistes de l'AEN, Paris, 1978)

£8.00   US$16.50   F66,00

Management, Stabilisation and Environmental Impact of Uranium Mill Tailings (Proceedings of the Albuquerque Seminar, United States, 1978)

Gestion, stabilisation et incidence sur l'environnement des résidus de traitement de l'uranium
(Compte rendu du Séminaire d'Albuquerque, États-Unis, 1978)

£9.80   US$20.00   F80,00

Exposure to Radiation from the Natural Radioactivity in Building Materials (Report by an NEA Group of Experts, 1979)

Exposition aux rayonnements due à la radioactivité naturelle des matériaux de construction
(Rapport établi par un Groupe d'experts de l'AEN, 1979)

Free on request — Gratuit sur demande

Marine Radioecology
(Proceedings of the Tokyo Seminar, 1979)

Radioécologie marine
(Compte rendu du Colloque de Tokyo, 1979)

£9.60   US$21.50   F86.00

Radiological Significance and Management of Tritium, Carbon-14, Krypton-85 and Iodine-129 arising from the Nuclear Fuel Cycle (Report by an NEA Group of Experts, 1980)

Importance radiologique et gestion des radionucléides : tritium, carbone-14, krypton-85 et iode-129, produits au cours du cycle du combustible nucléaire
(Rapport établi par un Groupe d'experts de l'AEN, 1980)

£8.40   US$19.00   F76,00

The Environmental and Biological Behaviour of Plutonium and Some Other Transuranium Elements (Report by an NEA Group of Experts, 1981)

Le comportement mésologique et biologique du plutonium et de certains autres éléments transuraniens (Rapport établi par un Groupe d'experts de l'AEN, 1981)

£4.60   US$10.00   F46,00

Uranium Mill Tailings Management (Proceedings of two Workshops) (in preparation)

La gestion des résidus de traitement de l'uranium
(Compte rendu de deux réunions de travail) (en préparation)

£ 7.20   US$ 16.00   F 72,00

RADIOACTIVE WASTE MANAGEMENT

GESTION DES DÉCHETS RADIOACTIFS

Objectives, Concepts and Strategies for the Management of Radioactive Waste Arising from Nuclear Power Programmes (Report by an NEA Group of Experts, 1977)

Objectifs, concepts et stratégies en matière de gestion des déchets radioactifs résultant des programmes nucléaires de puissance
(Rapport établi par un Groupe d'experts de l'AEN, 1977)

£8.50   US$17.50   F70,00

Treatment, Conditioning and Storage of Solid Alpha-Bearing Waste and Cladding Hulls
(Proceedings of the NEA/IAEA Technical Seminar, Paris, 1977)

Traitement, conditionnement et stockage des déchets solides alpha et des coques de dégainage
(Compte rendu du Séminaire technique AEN/AIEA, Paris, 1977)

£7.30   US$15.00   F60,00

Storage of Spent Fuel Elements
(Proceedings of the Madrid Seminar, 1978)

Stockage des éléments combustibles irradiés (Compte rendu du Séminaire de Madrid, 1978)

£7.30   US$15.00   F60,00

In Situ Heating Experiments in Geological Formations
(Proceedings of the Ludvika Seminar, Sweden, 1978)

Expériences de dégagement de chaleur in situ dans les formations géologiques
(Compte rendu du Séminaire de Ludvika, Suède, 1978)

£8.00   US$16.50   F66,00

Migration of Long-lived Radionuclides in the Geosphere
(Proceedings of the Brussels Workshop, 1979)

Migration des radionucléides à vie longue dans la géosphère
(Compte rendu de la réunion de travail de Bruxelles, 1979)

£8.30   US$17.00   F68,00

Low-Flow, Low-Permeability Measurements in Largely Impermeable Rocks
(Proceedings of the Paris Workshop, 1979)

Mesures des faibles écoulements et des faibles perméabilités dans des roches relativement imperméables
(Compte rendu de la réunion de travail de Paris, 1979)

£7.80   US$16.00   F64,00

On-Site Management of Power Reactor Wastes
(Proceedings of the Zurich Symposium, 1979)

Gestion des déchets en provenance des réacteurs de puissance sur le site de la centrale
(Compte rendu du Colloque de Zurich, 1979)

£11.00   US$22.50   F90,00

Recommended Operational Procedures for Sea Dumping of Radioactive Waste (1979)

Recommandations relatives aux procédures d'exécution des opérations d'immersion de déchets radioactifs en mer (1979)

Free on request — Gratuit sur demande

Guidelines for Sea Dumping Packages of Radioactive Waste
(Revised version, 1979)

Guide relatif aux conteneurs de déchets radioactifs destinés au rejet en mer
(Version révisée, 1979)

Free on request — Gratuit sur demande

Use of Argillaceous Materials for the Isolation of Radioactive Waste (Proceedings of the Paris Workshop, 1979)

Utilisation des matériaux argileux pour l'isolement des déchets radioactifs (Compte rendu de la Réunion de travail de Paris, 1979)

£7.60   US$17.00   F68,00

Review of the Continued Suitability of the Dumping Site for Radioactive Waste in the North-East Atlantic (1980)

Réévaluation de la validité du site d'immersion de déchets radioactifs dans la région nord-est de l'Atlantique (1980)

Free on request — Gratuit sur demande

Decommissioning Requirements in the Design of Nuclear Facilities (Proceedings of the NEA Specialist Meeting, Paris, 1980)

Déclassement des installations nucléaires : exigences à prendre en compte au stade de la conception (Compte rendu d'une réunion de spécialistes de l'AEN, Paris, 1980)

£7.80   US$17.50   F70,00

Borehole and Shaft Plugging (Proceedings of the Columbus Workshop, United States, 1980)

Colmatage des forages et des puits (Compte rendu de la réunion de travail de Columbus, États-Unis, 1980)

£12.00   US$30.00   F120,00

Radionucleide Release Scenarios for Geologic Repositories (Proceedings of the Paris Workshop, 1980)

Scénarios de libération des radionucléides à partir de dépôts situés dans les formations géologiques (Compte rendu de la réunion de travail de Paris, 1980)

£6.00   US$15.00   F60,00

Research and Environmental Surveillance Programme Related to Sea Disposal of Radioactive Waste (1981)

Programme de recherches et de surveillance du milieu lié à l'immersion de déchets radioactifs en mer (1981)

Free on request — Gratuit sur demande

Cutting Techniques as related to Decommissioning of Nuclear Facilities (Report by an NEA Group of Experts, 1981)

Techniques de découpe utilisées au cours du déclassement d'installations nucléaires (Rapport établi par un Groupe d'experts de l'AEN, 1981)

£3.00   US$7.50   F30.00

Decontamination Methods as related to Decommissioning of Nuclear Facilities (Report by an NEA Group of Experts, 1981)

Méthodes de décontamination relatives au déclassement des installations nucléaires (Rapport établi par un Groupe d'experts de l'AEN, 1981)

£2.80   US$7.00   F28,00

Siting of Radioactive Waste Repositories in Geological Formations
(Proceedings of the Paris Workshop, 1981)

Choix des sites des dépôts de déchets radioactifs dans les formations géologiques
(Compte rendu d'une réunion de travail de Paris, 1981)

£6.80   US$15.00   F68,00

Near-Field Phenomena in Geologic Repositories for Radioactive Waste
(Proceedings of the Seattle Workshop, United States, 1981)

Phénomènes en champ proche des dépôts de déchets radioactifs en formations géologiques
(Compte rendu de la réunion de travail de Seattle, Etats-Unis, 1981)

£11.00 $24.50   F110,00

● ● ●

SAFETY

Safety of Nuclear Ships
(Proceedings of the Hamburg Symposium, 1977)

SÛRETÉ

Sûreté des navires nucléaires
(Compte rendu du Symposium de Hambourg, 1977)

£17.00   US$35.00   F140,00

Nuclear Aerosols in Reactor Safety
(A State-of-the-Art Report by a Group of Experts, 1979)

Les aérosols nucléaires dans la sûreté des réacteurs
(Rapport sur l'état des connaissances établi par un Groupe d'Experts, 1979)

£8.30   US$18.75   F75,00

Plate Inspection Programme
(Report from the Plate Inspection Steering Committee — PISC — on the Ultrasonic Examination of Three Test Plates), 1980

Programme d'inspection des tôles
(Rapport du Comité de Direction sur l'inspection des tôles — PISC — sur l'examen par ultrasons de trois tôles d'essai au moyen de la procédure «PISC» basée sur le code ASME XI), 1980

£3.30   US$7.50   F30.00

Reference Seismic Ground Motions in Nuclear Safety Assessments
(A State-of-the-Art Report by a Group of Experts, 1980)

Les mouvements sismiques de référence du sol dans l'évaluation de la sûreté des installations nucléaires
(Rapport sur l'état des connaissances établi par un Groupe d'experts, 1980)

£7.00   US$16.00   F64,00

Nuclear Safety Research in the OECD Area. The Response to the Three Mile Island Accident (1980)

Les recherches en matière de sûreté nucléaire dans les pays de l'OCDE. L'adaptation des programmes à la suite de l'accident de Three Mile Island (1980)

£3.20   US$8.00   F32,00

Safety Aspects of Fuel Behaviour in Off-Normal and Accident Conditions
(Proceedings of the Specialist Meeting, Espoo, Finland, 1980)

Considérations de sûreté relatives au comportement du combustible dans des conditions anormales et accidentelles
(Compte rendu de la réunion de spécialistes, Espoo, Finlande, 1980)

£12.60 $28.00   F126,00

Safety of the Nuclear Fuel Cycle (A State-of-the-Art Report by a Group of Experts, 1981)

Sûreté du Cycle du Combustible Nucléaire
(Rapport sur l'état des connaissances établi par un Groupe d'Experts, 1981)

£6.60 $16.50   F66,00

LEGAL PUBLICATIONS

PUBLICATIONS JURIDIQUES

Convention on Third Party Liability in the Field of Nuclear Energy — incorporating the provisions of Additional Protocol of January 1964

Convention sur la responsabilité civile dans le domaine de l'énergie nucléaire — Texte incluant les dispositions du Protocole additionnel de janvier 1964

Free on request — Gratuit sur demande

Nuclear Legislation, Analytical Study: "Nuclear Third Party Liability" (revised version, 1976)

Législations nucléaires, étude analytique: "Responsabilité civile nucléaire" (version révisée, 1976)

£6.00   US$12.50   F50,00

Nuclear Legislation, Analytical Study: "Regulations governing the Transport of Radioactive Materials" (1980)

Législations nucléaires, étude analytique : "Réglementation relative au transport des matières radioactives" (1980)

£8.40   US$21.00   F84,00

Nuclear Law Bulletin
(Annual Subscription — two issues and supplements)

Bulletin de Droit Nucléaire
(Abonnement annuel — deux numéros et suppléments)

£6.00 $13.00   F60,00

Index of the first twenty five issues of the Nuclear Law Bulletin

Index des vingt-cinq premiers numéros du Bulletin de Droit Nucléaire

Description of Licensing Systems and Inspection of Nuclear Installation (1980)

Description du régime d'autorisation et d'inspection des installations nucléaires (1980)

£7.60   US$19.00   F76,00

NEA Statute

Statuts de l'AEN

Free on request — Gratuit sur demande

• • •

OECD SALES AGENTS
DÉPOSITAIRES DES PUBLICATIONS DE L'OCDE

ARGENTINA – ARGENTINE
Carlos Hirsch S.R.L., Florida 165, 4° Piso (Galería Guemes)
1333 BUENOS AIRES, Tel. 33.1787.2391 y 30.7122

AUSTRALIA – AUSTRALIE
Australia and New Zealand Book Company Pty, Ltd.,
10 Aquatic Drive, Frenchs Forest, N.S.W. 2086
P.O. Box 459, BROOKVALE, N.S.W. 2100

AUSTRIA – AUTRICHE
OECD Publications and Information Center
4 Simrockstrasse 5300 BONN. Tel. (0228) 21.60.45
Local Agent/Agent local :
Gerold and Co., Graben 31, WIEN 1. Tel. 52.22.35

BELGIUM – BELGIQUE
LCLS
35, avenue de Stalingrad, 1000 BRUXELLES. Tel. 02.512.89.74

BRAZIL – BRÉSIL
Mestre Jou S.A., Rua Guaipa 518,
Caixa Postal 24090, 05089 SAO PAULO 10. Tel. 261.1920
Rua Senador Dantas 19 s/205-6, RIO DE JANEIRO GB.
Tel. 232.07.32

CANADA
Renouf Publishing Company Limited,
2182 St. Catherine Street West,
MONTRÉAL, Que. H3H 1M7. Tel. (514)937.3519
OTTAWA, Ont. K1P 5A6, 61 Sparks Street

DENMARK – DANEMARK
Munksgaard Export and Subscription Service
35, Nørre Søgade
DK 1370 KØBENHAVN K. Tel. +45.1.12.85.70

FINLAND – FINLANDE
Akateeminen Kirjakauppa
Keskuskatu 1, 00100 HELSINKI 10. Tel. 65.11.22

FRANCE
Bureau des Publications de l'OCDE,
2 rue André-Pascal, 75775 PARIS CEDEX 16. Tel. (1) 524.81.67
Principal correspondant :
13602 AIX-EN-PROVENCE : Librairie de l'Université.
Tel. 26.18.08

GERMANY – ALLEMAGNE
OECD Publications and Information Center
4 Simrockstrasse 5300 BONN Tel. (0228) 21.60.45

GREECE – GRÈCE
Librairie Kauffmann, 28 rue du Stade,
ATHÈNES 132. Tel. 322.21.60

HONG-KONG
Government Information Services,
Publications/Sales Section, Baskerville House,
2/F., 22 Ice House Street

ICELAND – ISLANDE
Snaebjörn Jönsson and Co., h.f.,
Hafnarstraeti 4 and 9, P.O.B. 1131, REYKJAVIK.
Tel. 13133/14281/11936

INDIA – INDE
Oxford Book and Stationery Co. :
NEW DELHI-1, Scindia House. Tel. 45896
CALCUTTA 700016, 17 Park Street. Tel. 240832

INDONESIA – INDONÉSIE
PDIN-LIPI, P.O. Box 3065/JKT., JAKARTA, Tel. 583467

IRELAND – IRLANDE
TDC Publishers – Library Suppliers
12 North Frederick Street, DUBLIN 1 Tel. 744835-749677

ITALY – ITALIE
Libreria Commissionaria Sansoni :
Via Lamarmora 45, 50121 FIRENZE. Tel. 579751
Via Bartolini 29, 20155 MILANO. Tel. 365083
Sub-depositari :
Editrice e Libreria Herder,
Piazza Montecitorio 120, 00 186 ROMA. Tel. 6794628
Libreria Hoepli, Via Hoepli 5, 20121 MILANO. Tel. 865446
Libreria Lattes, Via Garibaldi 3, 10122 TORINO. Tel. 519274
La diffusione delle edizioni OCSE è inoltre assicurata dalle migliori
librerie nelle città più importanti.

JAPAN – JAPON
OECD Publications and Information Center,
Landic Akasaka Bldg., 2-3-4 Akasaka,
Minato-ku, TOKYO 107 Tel. 586.2016

KOREA – CORÉE
Pan Korea Book Corporation,
P.O. Box n° 101 Kwangwhamun, SÉOUL. Tel. 72.7369

LEBANON – LIBAN
Documenta Scientifica/Redico,
Edison Building, Bliss Street, P.O. Box 5641, BEIRUT.
Tel. 354429 – 344425

MALAYSIA – MALAISIE
and/et SINGAPORE - SINGAPOUR
University of Malaysia Co-operative Bookshop Ltd.
P.O. Box 1127, Jalan Pantai Baru
KUALA LUMPUR. Tel. 51425, 54058, 54361

THE NETHERLANDS – PAYS-BAS
Staatsuitgeverij
Verzendboekhandel Chr. Plantijnstraat 1
Postbus 20014
2500 EA S-GRAVENAGE. Tel. nr. 070.789911
Voor bestellingen: Tel. 070.789208

NEW ZEALAND – NOUVELLE-ZÉLANDE
Publications Section,
Government Printing Office Bookshops:
AUCKLAND: Retail Bookshop: 25 Rutland Street,
Mail Orders: 85 Beach Road, Private Bag C.P.O.
HAMILTON: Retail Ward Street,
Mail Orders, P.O. Box 857
WELLINGTON: Retail: Mulgrave Street (Head Office),
Cubacade World Trade Centre
Mail Orders: Private Bag
CHRISTCHURCH: Retail: 159 Hereford Street,
Mail Orders: Private Bag
DUNEDIN: Retail: Princes Street
Mail Order: P.O. Box 1104

NORWAY – NORVÈGE
J.G. TANUM A/S Karl Johansgate 43
P.O. Box 1177 Sentrum OSLO 1. Tel. (02) 80.12.60

PAKISTAN
Mirza Book Agency, 65 Shahrah Quaid-E-Azam, LAHORE 3.
Tel. 66839

PHILIPPINES
National Book Store, Inc.
Library Services Division, P.O. Box 1934, MANILA.
Tel. Nos. 49.43.06 to 09, 40.53.45, 49.45.12

PORTUGAL
Livraria Portugal, Rua do Carmo 70-74,
1117 LISBOA CODEX. Tel. 360582/3

SPAIN – ESPAGNE
Mundi-Prensa Libros, S.A.
Castelló 37, Apartado 1223, MADRID-1. Tel. 275.46.55
Libreria Bastinos, Pelayo 52, BARCELONA 1. Tel. 222.06.00

SWEDEN – SUÈDE
AB CE Fritzes Kungl Hovbokhandel,
Box 16 356, S 103 27 STH, Regeringsgatan 12,
DS STOCKHOLM. Tel. 08/23.89.00

SWITZERLAND – SUISSE
OECD Publications and Information Center
4 Simrockstrasse 5300 BONN. Tel. (0228) 21.60.45
Local Agents/Agents locaux
Librairie Payot, 6 rue Grenus, 1211 GENÈVE 11. Tel. 022.31.89.50
Freihofer A.G., Weinbergstr. 109, CH-8006 ZÜRICH.
Tel. 01.3634282

THAILAND – THAILANDE
Suksit Siam Co., Ltd., 1715 Rama IV Rd,
Samyan, BANGKOK 5. Tel. 2511630

TURKEY – TURQUIE
Kültur Yayinlari Is-Türk Ltd. Sti.
Atatürk Bulvari No : 77/B
KIZILAY/ANKARA. Tel. 17 02 66
Dolmabahce Cad. No : 29
BESIKTAS/ISTANBUL. Tel. 60 71 88

UNITED KINGDOM – ROYAUME-UNI
H.M. Stationery Office, P.O.B. 569,
LONDON SE1   9NH. Tel. 01.928.6977, Ext. 410 or
49 High Holborn, LONDON WC1V 6 HB (personal callers)
Branches at: EDINBURGH, BIRMINGHAM, BRISTOL,
MANCHESTER, CARDIFF, BELFAST.

UNITED STATES OF AMERICA – ÉTATS-UNIS
OECD Publications and Information Center, Suite 1207,
1750 Pennsylvania Ave., N.W. WASHINGTON, D.C.20006 – 4582
Tel. (202) 724.1857

VENEZUELA
Libreria del Este, Avda. F. Miranda 52, Edificio Galipan,
CARACAS 106. Tel. 32.23.01/33.26.04/33.24.73

YUGOSLAVIA – YOUGOSLAVIE
Jugoslovenska Knjiga, Terazije 27, P.O.B. 36, BEOGRAD.
Tel. 621.992

Les commandes provenant de pays où l'OCDE n'a pas encore désigné de dépositaire peuvent être adressées à :
OCDE, Bureau des Publications, 2, rue André-Pascal, 75775 PARIS CEDEX 16.

Orders and inquiries from countries where sales agents have not yet been appointed may be sent to:
OECD, Publications Office, 2 rue André-Pascal, 75775 PARIS CEDEX 16. *

PUBLICATIONS DE L'OCDE, 2, rue André-Pascal, 75775 PARIS CEDEX 16 - N° 42181 1982
IMPRIMÉ EN FRANCE
(66 82 02 3) ISBN 92-64-02288-0